ROUTLEDGE LIBRARY EDITIONS:
ECONOMIC GEOGRAPHY

Volume 10

AN INDUSTRIAL GEOGRAPHY OF
THE NETHERLANDS

AN INDUSTRIAL GEOGRAPHY OF THE NETHERLANDS

An International Perspective

MARC DE SMIDT AND EGBERT WEVER

Translated by
NANCY SMYTH VAN WEESEP

Routledge
Taylor & Francis Group

LONDON AND NEW YORK

First published by Routledge in 1990

This edition first published in 2015
by Routledge
2 Park Square, Milton Park, Abingdon, Oxon, OX14 4RN

and by Routledge
711 Third Avenue, New York, NY 10017

Routledge is an imprint of the Taylor & Francis Group, an informa business

British Library Cataloguing in Publication Data
A catalogue record for this book is available from the British Library

ISBN: 978-1-138-85764-3 (Set)
eISBN: 978-1-315-71580-3 (Set)
ISBN: 978-1-138-88475-5 (Volume 10)
eISBN: 978-1-315-71568-1 (Volume 10)
Pb ISBN: 978-1-138-88508-0 (Volume 10)

Publisher's Note
The publisher has gone to great lengths to ensure the quality of this reprint but points out that some imperfections in the original copies may be apparent.

Disclaimer
The publisher has made every effort to trace copyright holders and would welcome correspondence from those they have been unable to trace.

AN INDUSTRIAL GEOGRAPHY OF THE NETHERLANDS

An
International Perspective

Marc de Smidt
and
Egbert Wever

Translated by Nancy Smyth Van Weesep

Routledge
London and New York

Dutch edition *De Nederlandse Industrie. Positie, spreiding en struktuur.*
Published by Van Gorcum, Assen/Maastricht 1987.
Published in 1990
by Routledge
11 New Fetter Lane, London EC4P 4EE

Simultaneously published in the USA and Canada
by Routledge
a division of Routledge, Chapman and Hall, Inc.
29 West 35th Street, New York, NY 10001

Printed and bound in Great Britain by Mackays of Chatham PLC, Kent

British Library Cataloguing in Publication Data

Wever, Egbert
 An industrial geography of the Netherlands: an
 international perspective. – (Industrial geography
 series)
 I. Title II. Smidt, Marc de, *1941–* III. Series
 338.09492

 ISBN 0-415-00153-6

Library of Congress Cataloging in Publication Data
also applied for

Contents

List of figures

viii

List of tables

Acknowledgements

The authors wish to thank the help received from the trans-
lator, Nancy Smyth Van Weesep. In addition they would like
to acknowledge the assistance of the secretaries within the
Utrecht Department of Geography: Wilma Admiraal, Marianne
Hamel and Ilse van der Lek, as well as the Cartographic
Laboratory of the Department and the lay-out assistance of
Gerard van Betlehem, Margriet Tiemeijer and Theo Woudstra.

1 Introduction

1.1 Manufacturing and the Social Context

Industrial activities have almost always received a good deal of attention. Perhaps this is because modern industry provided the foundation for today's pattern of urbanization. Indeed, the emergence of our cities is intertwined with the transition from traditional cottage industries to production in factories. It is also possible that the attention for industrial activities is brought about by the fact that they are usually highly concentrated in space. Through inertia, industrial estates and industrial zones maintain their reputation, if not their characteristic appearance, even if they lose their manufacturing function. A third possible factor may be found in the locational preferences of industrial firms. Modern industry in particular tends to be footloose and thus can exercise considerable freedom in selecting a location. Consequently, regional policy, which attempts to balance the distribution of both jobs and the labor force, was for a long time essentially industrialization policy. This explains why administrators and politicians focus on industrial firms in their attempt to draw businesses to their communities. A fourth factor could be the presence in the Netherlands of a number of corporations of international renown and importance. Among these are certainly Shell, Unilever and Philips, and probably firms like Hoogovens, AKZO, and Fokker.

Despite this attention, industrial activities are not always appreciated by society as a whole. The reconstruction period of the national economy after 1945 was a glorious era for Dutch industry. But this changed in the 1960s and 1970s. The continued economic expansion then brought a number of negative effects to light, certainly in specific regions: traffic congestion, degradation of the environment, insalubrious housing conditions, greater commuting distances, etc. There was increasing talk of selective growth and the necessity to alleviate the pressure on De Randstad as a reaction to these problems. For many it was clear that

Fig. 1.1   The Netherlands: topography

manufacturing was the culprit that caused these negative
effects. Especially in the Randstad area, people thought
they would be better off without industry. This negative
attitude was fueled by the decline in the number of jobs in
manufacturing after the mid-1960s, a consequence of rationalization and automation operations. At the same time the
stature of the service sector was enhanced because it provided increasing numbers of jobs and it did not inflict
visual damage on the environment.

By the end of the 1970s, however, the attention for manufacturing rose again, and with it came greater appreciation. Undoubtedly, this change is related to the economic recession. But another important factor was the publication of a report on the state and future of manufacturing in the Netherlands by the WRR, the scientific advisory board of the government. The report gave a gloomy picture of the strength of Dutch industry in the face of international competition. At the same time, it offered a number of suggestions for improvement. The report suggested that the only effective course of action would be to upgrade the industrial mix. The Netherlands should specialize in those products and sectors for which highly trained labor and specialized know-how are necessary. Obviously, such a change can only take place in the long term. Within a short span of time, industrial structures are rather inflexible, a fact to which many older industrial estates bear witness.

The activities of the Commission on the Progress of the Industrialization Policy (known as the 'Wagner Commission', after its chairman), which provided a follow-up to the report on the state and future of manufacturing in the Netherlands, further heightened the attention for industry. Concepts that were used frequently in the deliberations of this commission, such as 'successful activities', 'high-tech activities', 'product and process innovations' are mostly linked to industrial activities.

These developments in attitudes with regard to manufacturing provide the context for this study of manufacturing in the Netherlands. Industry has clearly lost much of the predominance it had in the economy of the early 1960s. But even today, it still comprises about 20 percent of the national labor volume. Its share in the total value added in the private sector is also roughly 20 percent. Its position with respect to investments in capital goods is somewhat weaker (15 percent), but manufactured products (including natural gas) account for a much more important share of total exports: well over 70 percent!

1.2 Manufacturing: Position, Distribution, Structure

Re-industrialization formed the basis for the reconstruction of the Dutch economy after 1945. In Chapter 2 the role and function of manufacturing in the reconstruction is discussed in detail. This discussion focuses on the industrialization debate that was carried on after 1945. The changing industrial structure and industrial policy are traced up to 1963, at which time the Netherlands appeared to be a mature industrial nation and the process of restructuring was initiated. The key position of manufacturing in reconstruction may seem surprising, in that the Netherlands was an industrial late bloomer. This warrants a brief treatment in Chapter 2 of the rise and development of manufacturing in the period prior to 1940. Although the emphasis is on the position of Dutch manufacturing within a wider societal context, some attention is also given to the distribution and the structure of industry.

Chapters 3 and 4 are closely related to Chapter 2. First, Chapter 3 treats the changes in the sectoral structure. Each period has its own particular sectors of growth and contraction. On the one hand, this is because consumer preferences change. The life cycle of manufactured products is brought up in this connection: for example, the replacement of black and white televisions by color sets. On the other hand, this is related to possible changes in international competitive positions. The expansion of certain industrial activities in one region or country can lead to contraction elsewhere. The rise of countries such as Japan and, later on, South Korea as shipbuilding nations and the dismantling of this sector in countries such as the Netherlands and Sweden is an example of this reciprocal relationship. Over time, this process generates strong fluctuations in the structure of industry. The dynamics of industry are discussed in depth in Chapter 3. After the successful industrialization after 1945, the international competitive position of Dutch industry weakened. The threat of de-industrialization provoked attempts to re-industrialize the economy. Re-industrialization was conceived as a way to make use of the country's comparative advantages. This re-industrialization debate, in which Van der Zwan and Wagner took part, is also treated in Chapter 3.

Insight into sectoral dynamics is important for a thorough understanding of how the Dutch economy was built up and developed. Many growth and contraction activities are regionally concentrated. Classic examples are the leather goods and footwear industries in the area known as Langstraat in the province of Noord-Brabant and the textile industry in the region of Twente. Sectoral dynamics thus induce expansion and contraction and thereby often generate spatial dynamics. The distribution of industrial activity in the Netherlands is the central topic of Chapter 4. This discussion also deals with the importance of industry for the various regions. Since regional specializations tend to persist in the mental maps that people form of the Netherlands (e. g. the image of Twente and Langstraat), this chapter goes back to the period around 1930. At that time, the foundation was laid for a spatial pattern of specialization that was maintained for a long time afterwards.

The aspect of distribution is also the focus of Chapter 5, but in this case from the perspective of policy. Previous chapters dealt with industrialization policy at the national

1. Oost-Groningen
2. Delfzijl e.o.
3. Overig Groningen
4. Noord-Friesland
5. Zuidwest-Friesland
6. Zuidoost-Friesland
7. Noord-Drenthe
8. Zuidoost-Drenthe
9. Zuidwest-Drenthe
10. Noord-Overijssel
11. Zuidwest-Overijssel
12. Twente
13. Veluwe
14. Achterhoek
15. Agglomeraties Arnhem
    en Nijmegen
16. Zuidwest-Gelderland
17. Utrecht
18. Kop van Noord-Holland
19. Alkmaar e.o.
20. IJmond
21. Agglomeratie Haarlem
22. Zaanstreek
23. Groot-Amsterdam
24. Het Gooi en Vechtstreek
25. Agglomeratie Leiden en
    Bollenstreek
26. Agglomeratie 's-Gravenhage
27. Delft en Westland
28. Oost Zuid-Holland
29. Groot Rijnmond
30. Zuidoost Zuid-Holland
31. Zeeuwsch-Vlaanderen
32. Overig Zeeland
33. West Noord-Brabant
34. Midden Noord-Brabant
35. Noordoost Noord-Brabant
36. Zuidoost Noord-Brabant
37. Noord-Limburg
38. Midden-Limburg
39. Zuid-Limburg
40. Zuidelijke IJsselmeerpolders

Fig. 1.2   The Netherlands: regions (COROP)

level, so this chapter concentrates on policy at the regional level. This regional policy, formulated at the end of the
1940s, is characterized by three essential changes. In the
first place, it shifted from an almost exclusively industrial policy to one in which other economic sectors became
increasingly important. Secondly, it moved away from incentives and toward development policy. In the third place,
whereas policy had been the prerogative of government, a
greater role was carved out for the regions to implement
their own policy. Chapter 5 treats each of these shifts.

5

By definition, government policy is geared to a more or less distant future. Chapter 6 adopts that perspective as well. The steps toward renewal in industry and the opportunities for the future of various sectors are central to the discussion of the future. This chapter, which in many cases runs parallel to the activities of the 'Wagner Commission', elaborates on the distribution of high-potential activities in the context of innovative enterprise, innovative production environments and new business initiatives. This kind of information gives an indication of how Dutch industry will be distributed in the near future. The findings deviate somewhat from the commonly held spatial image of the Dutch economy. That image may be roughly characterized as De Randstad versus the rest of the Netherlands. The main thrust of this discussion of industry in the Netherlands is explicated most clearly in Chapter 6. The essence of the issues treated in this book may be summarized in terms of three processes: deconcentration, the dissolution of regional differentiation and the expansion of the Dutch urban field.

In each of the first six chapters, though most explicitly in Chapter 3, the open economy of the Netherlands is constantly in the background. This open character defines the constraints and the opportunities of the Dutch economy to such a high degree that it was deemed necessary to give this background extra attention. Chapter 7 therefore elaborates on the multinationalization of business and on the foreign firms that have located in the Netherlands. The aspects of position, distribution and structure are central to this discussion. Case studies are used to present the industrial developments around Schiphol and in the port of Rotterdam, which together form the core of the Netherlands' international role as the 'Gateway to Europe'. Chapter 8 deals with the largest Dutch manufacturing firms in terms of their origin, growth, restructuring and location.

Chapter 9, finally, summarizes the most salient findings. The concept of regional-economic potential provides a medium through which to explain the observation that spatial imagery no longer corresponds to spatial reality.

Throughout the book readers not well acquainted with the topography and toponomy of the Netherlands will encounter many names of cities, regions and provinces that may be unfamiliar to them. Figures 1.1 and 1.2 may help them place these names on a map.

2 The Industrialization Process in Perspective

2.1 The Need to Industrialize

In 1947, when the American Secretary of State Marshall addressed a group at Harvard on the necessity of launching a grand-scale European Recovery Program, the economy of the Netherlands was on the brink of collapse. In the two years since the end of the Second World War, work to repair the staggering war damage had hardly begun. Half of the domestic capital and 40 percent of the total production capacity had been lost. The energy and transportation infrastructures were in even worse shape. Food shortages undermined 'our power to work' which, according to Prime Minister Schermerhorn, 'is virtually all we have left' (Van der Linden 1985: 139). By 1947, the national reserve of dollars, which was being used to finance the import of goods to repair the war damage, was diminishing alarmingly, despite the liquidation of foreign assets. This depletion of reserves threatened to jeopardize the reconstruction of industry.

The problem of recovery and reconstruction was not a temporary one; the situation disclosed a number of structural problems. The problem that raised most concern was the employment situation. The memory of half a million unemployed at the trough of the Depression was still fresh. In 1936, one-fifth of the labor force was unemployed, not to mention the immense number of underemployed.

The underlying problem was the position of the Netherlands as an industrial country, an issue that has not been fully resolved in the decades since the war. Attention for this issue was revived by the publication of a report on the present and future role of manufacturing in the Netherlands (WRR 1980). This does not make the early postwar experience less interesting. On the contrary, the debate on industrialization, carried on during the first decade after the Second World War, offers insight in the fundamental obstacles facing the country at that time. But in following chapters, the discussion of current industrialization problems will not only throw light on how the problems were previously resolved but will also clarify whether the issue of industrialization is still relevant. This presumes some understanding of the Netherlands as an industrial late bloomer, engaged until 1963 in adjusting to the European industrial pattern.

The Bleak Outlook in 1945

After 1945, industrial recovery and economic growth were not the only problems; population growth had become another important issue. Around 1937 the previously declining birth-rate reversed; by 1946 the growth was a veritable explosion. In 1949, when the Dutch population reached the ten million mark, 60 percent of the respondents to a survey expressed deep concern about the consequences of the rapid population growth, such as unemployment, excessive population density, scarcity of food, and shortage of housing (Heeren 1985).

Employment had to be created for the vigorously growing labor force, primarily through industrialization, because agriculture was bound to shed labor. Emigration to countries such as Canada and Australia was seen as a recourse (annexation of German border areas was even considered as a way to augment agricultural production and resource exploitation!). The number of emigrants actually reached 48,690 in 1952. This way of easing the population pressure enjoyed strong governmental support.

Besides unemployment, the so-called external account was cause for concern. The balance of payments in the Netherlands traditionally consisted of a balance-of-trade deficit (imports exceeding exports) and a favorable balance in services and capital. The latter, however, had been severely disrupted by the war. Numerous relations had to be redeveloped after 1945 and many foreign assets were liquidated to support the cause of economic recovery. The loss of the Dutch East Indies not only brought on the demise of Amsterdam as a hub of colonial trade; a protected export market was also lost. The three-way trade pattern disintegrated: export of resources from the East Indies to the United States had provided the Netherlands with the currency to import (industrial) resources and capital goods from the dollar areas (Van der Linden 1985). The decolonization process forced the Netherlands to acquire new sources of foreign exchange. Thereby, the traditional trade deficit could no longer be tolerated, despite the need to compensate for the domestic lack of natural resources to supply the manufacturing sector.

The Industrialization Debate

The debate about the necessity for a grand-scale industrialization program, due to the employment squeeze and the unfavorable balance of trade, flared up in 1947. According to Kohnstamm (1947), the pillars of the industrial strategy should be the basic metal and chemical industries, supported by research, in order to build an export-oriented, high-quality industry. This, however, was easier said than done. The labor force lacked the manufacturing skills, and government and the business community lacked the appropriate attitude. The industrialization offensive was to include consciousness-raising among the Dutch public and labor force in regard to the quality of the industrial products. No clear consensus was reached on this tactic, however.

The industrialization debate should be seen in a wider context. Which model of society was being pursued? 'The Plan for Labor shall free the Netherlands from the heavy burden of unemployment.' This rhetoric was voiced in 1935 by the socialist movement, which advocated supporting the level of demand (an anti-cyclical policy, like that propagated by Keynes) and controlling the supply (imposing discipline on business interests) to help lift the economy out of its deep depression. Governmental steering, with a centrally guided wage and price policy, sharply progressive taxation and introduction of a system of social security (including child support payments, unemployment benefits and, later, social security pensions for the aged) formed the main policy lines of the postwar 'Catholic-Red' coalition governments (de Liagre Böhl *et al.* 1981, 76 ff).

Although the goal of the Plan for Labor, 'security at a decent standard of living', was not attained for a long time, this policy paved the way for the first steps in that direction. There was no political majority in favor of a centralized economy, which would entail nationalization of key sectors (mines, basic industry, banks and insurance companies, etc.). Only the National Bank of the Netherlands was placed under governmental control. Free-market capitalism was not reinstituted intact, however: 'although ... intervention was not deemed desirable, the instruments of price and wage control in particular fulfilled an important policy role over a long period, primarily in fighting inflation, improving the international competitive position and distributing income as fairly as possible' (Van der Linden 1985: 85). The goal of coordinating business activity was not only pursued in the socialistic movement but also in religiously aligned (Roman Catholic and Protestant) circles.

Industrialization Policy

After 1945, industrialization was seen as the paramount challenge facing the nation. The (first) Memorandum on the

Industrialization of the Netherlands, issued in 1949 by the
Secretary of Economic Affairs Van den Brink, set the goal of
creating 215,000 new manufacturing jobs in four years.
Remarkably, 155,000 of these jobs were to stem from new
investment, especially in the metal, chemical and construction industries. While the competitor Germany was temporar-
ily absent from the field, metal and chemical companies had
a rare opportunity to take root. In addition, 60,000 jobs
were to be generated through a better utilization of existing investments (e.g. by working in shifts), particularly
in the textile, food and metal industries. An incipient

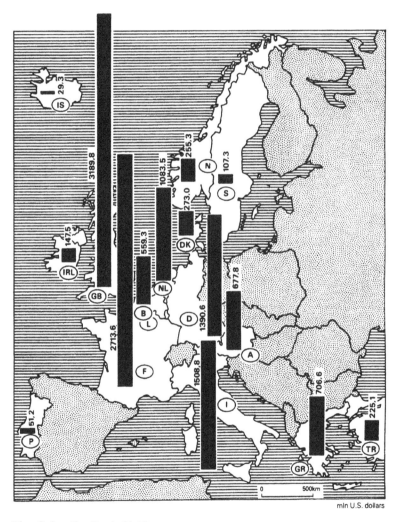

Fig. 2.1   The Marshall Plan

division between modern and traditional sectors seems to have manifested itself at that time. Investment was promoted with the support of the European Recovery Program (the Foreign Assistance Act, that implemented the so-called Marshall Plan from 1948 to 1952) and guaranteed by the World Bank. The volume of Marshall aid to the Netherlands was remarkably high in relation to the size of the economy (Fig. 2.1). The embargo on the import of capital goods was lifted. At the same time, adequate generation of energy was given high priority. However, despite the Marshall aid, the necessity to import resources and capital goods implied that the export position had to be structurally bolstered. This revived an old dilemma. Import substitution was, of course, only a limited option for a country with a small domestic market and a narrow scope for protection. This is a major reason why the Netherlands was the last stronghold of free trade during the Depression, yielding to the reality of worldwide protectionism only in the mid-1930s. Yet despite this turnabout, due to its location, the country would continue to take advantage of the growth of world trade. Agriculture and international services not only strengthened the export position for manufacturing; these sectors also helped balance the external account. The Netherlands was able to carry out these reforms because of the devaluation of the Dutch guilder - along with other European currencies - by 32 percent (!) in relation to the US dollar.

The Netherlands has been occasionally reproached internationally for having set up an overly ambitious industrialization program, but the government discounted the criticism by referring to population developments and the balance of payments (Van der Linden 1985). The government actually preferred labor-intensive, high-quality industrial activities, but understood that the requisite industrial tradition was not yet available (Wijers 1982). A second goal was to strengthen basic industries in order to fill fundamental gaps in the industrial structure and to radiate multiplier effects on other industries. Steel mills, oil refineries and the chemical plants of the National Mines (DSM) could therefore count on receiving credit and guarantees (for a detailed review, see Weisglas 1952).

The achievements of the period 1948-1952, construed to be the first stage of industrialization by Wijers (1982), formed the basis for the development of the Netherlands into a modern industrial country. The die was cast, but the prospects were bleak. 'It is far from probable that the expansion of employment in the coming years will keep up with the pace of population growth', in the opinion of Steigenga (1949: 71), and he was definitely not alone in this view. However, when Secretary of Economic Affairs Zijlstra presented the Fourth Memorandum on Industrialization in 1953, the goals formulated in the first memorandum had already been surpassed. In three years' time, production had increased by 17 percent and employment by 5 percent, indicating that labor productivity had increased considerably. Moreover, exports had risen by 69 percent, in contrast to imports, which had risen 'only' 19 percent. A clear trend toward industrial recovery is described by the curve of industrial production volume (Fig. 2.2).

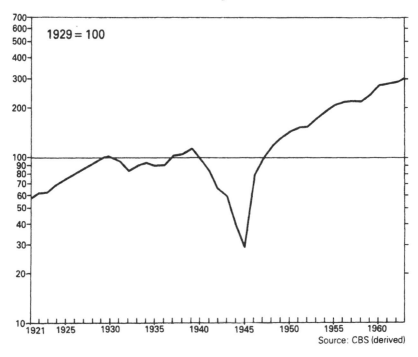

Source: CBS (derived)

Fig. 2.2   Production indices for the manufacturing
            sector, 1921-1963

Surprisingly, the discussion on the imperative of industrialization entirely ignored the opportunities in the tertiary sector. The role of services is demonstrated by two telling facts: the principal contribution to balancing the external account came from that side (the trade deficit would continue until 1952); and the contribution of the tertiary sector to employment growth, which had been pegged at 42,000, turned out to be much higher (85,000) (De Liagre Böhl *et al.* 1981: 228).

Altogether, the beginning of the postwar industrialization process was a clear success. In fact, it was the harbinger of an economic revival of wider scope.

2.2 The Netherlands: An Industrial Late Bloomer

The industrialization debate sketched above gives rise to the question whether there had been a break in the economic development of the Netherlands. The long climb upwards in the 1950s is sometimes seen as a breakthrough, an upswing of the long wave, which was firmly guided by government for the first time. The phenomenon of the long wave will be discussed first, as this shows the technological-economic background of the industrialization process. Then the question of why the Netherlands is an industrial late bloomer is answered. When exactly did the industrial breakthrough occur, and why did its character differ from developments elsewhere in Western Europe? By first reviewing the past, we can put the postwar industrialization debate into a better perspective.

<u>Innovations, the 'Leading Sector' and the 'Long Wave'</u>

It is believed that industrialization can only succeed when there is economic growth across the board. Therefore it has often been proposed that government should provide the conditions favorable for industrialization, such as an adequate infrastructure, elementary education and some degree of population concentration. Although these notions about balanced growth may be valuable, a number of authoritative theorists point out the importance of concentrating the effort within one or only a few parts of the economy, allowing it to surge forward and make the intended breakthrough to growth.

Hirschmann (1958) distinguishes a strategic sector that can work as a catalyst to bring on a new stage of growth in economic development. Sectors are not dealt with simultaneously but consecutively, due to the consequences of intersectoral relations (for example by supplies and expenditures).

Rostow (1963) elaborated this point of view, placed his theoretical considerations in a historical perspective and argued that technical innovations in the production process have had a decisive impact on the transition to a new stage of growth. He differentiated five stages: the 'traditional society', 'establishing the preconditions for take-off', the 'take-off', the 'drive to maturity' with self-sustained growth and, finally, society's arrival at the 'age of high mass consumption'. During the take-off, the leading sector is manifested: it is the push in the process of economic growth from which impulses are radiated out toward other branches of industry. An example is the complex formed by steel mills and the railway network. Of course, there is a specific leading sector peculiar to every historic period.

13

Schumpeter (1939) had already indicated that innovations were responsible for new impulses in economic growth. His concept of innovation encompassed more than technical inventions, and he emphasized the role of the entrepreneur in the organizational renewal of the process of social reproduction. It is the business community that brings about a 'Durchsetzung neuer Kombinationen', an introduction to new combinations of existing ideas. The work of Schumpeter has been revived in the last few years by the discussion of the impulse that technological innovation could help to overcome a drawn-out economic recession. His ideas on the relation between a rapid succession of innovations and renewed economic growth were expounded in the 'theory of the long wave'.

The 'long wave' theory goes back to Kondratieff, a Russian economist who attempted to penetrate the stages of capitalism (Van Duijn 1979, 1983). Pursuing this further, Schumpeter differentiated the long wave (40 to 60 years in duration) in four stages: prosperity, recession, depression and recovery. The start of a new Kondratieff cycle lies where the basic innovations (for example, television, transistors, synthetic fibers) have an impact on economic growth. Eventually a state of saturation or even overcapacity is reached. A specific moment arrives, i.e. a depression, when old innovations no longer offer much perspective, which Gerhard Mensch (1975) termed 'das technologische Patt', a technological stalemate. Recovery may occur when the overcapacity is cut back, investments increase and new basic innovations recapture the confidence of the market.

The peaks and troughs of a 'long wave' could coincide with the much shorter conjunctural cycle of Kitchin (with a duration of three to five years) and of Juglar (seven to eleven years). The Kondratieff cycle is less concerned with investment in stock or in machines than with long-life capital: investment in infrastructure and industrial complexes. The first long cycle was identified by Kuznets (1953) as the textiles phase of the Industrial Revolution (1787-1842), with the steam engine as its motor; the second embraced the so-called railway phase (with Bessemer steel as the basic innovation) and ended with the introduction of chemicals and electro-technology (1843-1897); the third followed suit and demonstrated applications of the internal combustion engine (automobile) and the electric motor (1898-1949). The fourth Kondratieff cycle commenced in 1949 and its key industries were electronics, petrochemicals and aeronautics. Gerhard Mensch (1975) predicted that the fifth Kondratieff cycle would arrive riding the crest of the wave made by the chip and bio-technology.

It would be interesting to confront the theories of Hirschmann, Rostow and Schumpeter with the stages in the industrialization process of the Netherlands, but it would also be difficult. The location of the Netherlands, surrounded by the industrial countries of Western Europe, each of which has its own industrial history, and its extremely open (and small) economy make it hard to trace the strategic sector of Hirschmann, the stages of Rostow and the Kondratieff cycles clearly.

The spatial dimension of these cycles of economic growth is even more difficult to outline. The theory of Myrdal (1957) assumes that a strong growth impulse is an inseparable accompaniment to an economy with a geographic concentration of activities; but growth then aggravates the center-periphery opposition. Congestion will eventually arise in the center, and fringe development will occur, although this will not extend to the periphery where the 'backwash' still persists. Pred (1977) pointed out how cumulative growth may help the evolution of an urban system, due to the intensity of contacts, amenities and the homogeneous labor market offered by cities.

The theories mentioned above provide a base on which to evaluate the industrialization of the Netherlands in the long term. In Section 2.4 the developments brought to the fore in Sections 2.2. and 2.3 will be placed in this theoretical context.

Industrial Breakthroughs and Renewal

The late arrival of the Netherlands on the industrial scene has been explained in the literature in at least three contexts: institutional-psychological, macro-economic and economic-geographical. These explanations are complementary in many ways.

The institutional-psychological explanation is the oldest. It refers to a 'negotiated environment' (Van der Haas 1967). Investors in the Netherlands were typically oriented towards cultivation in the tropics and towards international trade. It was not surprising to find a tradition of *laissez-faire* trade policy in a country with a limited home market and a favorable transport location, surrounded by countries whose access to raw materials had allowed them to industrialize early; these raw materials were (initially) unavailable in the Netherlands. On another level, the absence of the spirit of enterprise and innovation in the first half of the nineteenth century was widely criticized, even in literary works. Investors did not tend to form venture capital; when they did, they preferred to invest abroad.

The macro-economic approach, in contrast, is based on factor cost differentials. It was hard to accumulate capital in the Netherlands; labor was also costly, as comparison

15

with Belgium shows (Mokyr 1974). The low productivity of labor and the lack of skilled labor in manufacturing (many German tradesmen were recruited) in the Netherlands was a common complaint. This macro-economic explanation may be elaborated by placing it in an economic-geographical context. From a regional perspective, the labor costs in the textile industry in the Dutch province of Noord-Brabant proved to be lower than in the Belgian city of Liege, weakening the thrust of Mokyr's argument (Bos 1976). But the major obstacle for the Dutch textile industry areas was the absence of coalmines nearby. Even though the textile industry was strongly labor-oriented, it still had to haul coal to fuel the steam engines. The transport cost was so high that it was not compensated by the lower cost of labor, at comparable levels of productivity, found in areas where coal was more accessible. In Weber's terminology, the critical isodapane is exceeded in such a region.

This demonstrates the importance of an economic-geographical explanation premised on the regional cost differential of locational factors. It is also clear that important changes took place during the nineteenth century in the economic-geographical position of the Netherlands, in the relations between the trade center (the West) and the peripheral provinces, and in the relations of the country with its hinterland (particularly Germany), as documented by the economist-historian Bos (1976: 195). He points out that the the industrial ascent of Germany propelled the Netherlands into a central position instead of allowing it to remain peripheral in regard to the already industrialized countries of England and Belgium. Was this only to the advantage of certain ports? The Dutch industrial breakthrough should definitely not be ascribed to the international function of its ports alone. Shipbuilding and its associated machinery assembly and repair were built on that foundation; large interrelated industrial complexes were not, however. At that time, industrial complexes depended on the working of coal deposits, yet the exploitation of Dutch coalmines only commenced in the 1890s and continued at a very slow pace. The late industrialization of the Netherlands lacked a true leading sector, which, according to Rostow, characterizes the stage of take-off and the subsequent stage of self-sustained growth (De Jonge 1968). In the 1880s a national system of navigable waterways and a widely branched railway network became available; this infrastructure played a vital role in paving the way for industrialization. By the time the depression of the 1880s was superseded by the upswing of the third Kondratieff cycle in the early 1890s, the Dutch business community was ready to move with it. The late arrival of industrialization had one clear advantage, however. Activities based on new technology could be initiated without having to restructure the economy.

Although the true breakthrough only came in the 1890s, earlier attempts had been made to foster industrialization. New enterprises had been established, series and mass production had been organized, and a new legal construction for incorporation (the 'NV') allowed small businesses to take financial risks that did not entail personal liability. The breakthrough was characterized by a structural expansion of the economy. A new orientation towards capital goods industry and innovative technological areas such as electrical and chemical engineering was introduced, alongside the sectors of food and kindred products (including tobacco and alcoholic beverages) and textiles, which were already in place. The 'old' sectors were already well versed in mechanically powered factory production. Recent insights emphasize the important part played in the period around 1850 by a number of industrial groups concentrated in certain regions, for example the textile industry in Twente and Tilburg and the metal industry in certain cities in the provinces of Holland (Griffiths 1979). But entrepreneurs who dared to make 'Neue Kombinationen', as Schumpeter called innovation, were a rare breed in the first half of the nineteenth century (Zappey, in Van Stuijvenberg 1978: 218; see also De Jonge 1978).

Table 2.1 Employment in manufacturing, by sector, 1849-1909.

Branches	number of jobs (x 1000)			percentages in manufacturing			share in growth	
	1849	1889	1909	1849	1889	1909	1849/ 1889	1889/ 1909
I food & kindred products	32	72	116	11	15	16	22	19
II textiles, apparel, footwear, leather goods	126	140	177	42	29	25	8	15
III lumber & furniture	28	59	75	9	12	10	17	7
IV building materials & construction	65	128	180	22	27	25	35	22
V metals	33	57	108	11	12	15	13	22
VI new branches (chemicals, paper, printing, public utilities)	8	20	49	3	4	5	7	9
VII other branches (unclassified)	7	3	11	2	1	3	1	6
total	299	481	716	100	100	99	101	100

Source: Raw data derived from J.A. De Jonge 1968, 228-29.

Table 2.1 shows the development in certain industrial sectors before and after the industrial breakthrough. The territory gained by the 'new' sectors (chemicals, paper, printing) was still small after 1889; their role only became important in the second and third decades of the twentieth century. For the time being, the expansion of the 'traditional' sectors (food and kindred products, textiles) was still considerable, even after the industrial breakthrough. It should be kept in mind that the size of enterprises increased in these sectors and many crafts enterprises disappeared (see De Jonge 1968 for a more detailed analysis).

Functional and Spatial Concentration: The Rise of the Large Corporation

The industrial-geographic image of the Netherlands around 1930 is a mosaic of traditional regional specializations (Chapter 4, Fig. 4.1). Although it was then 'common knowledge that Noord-Brabant's Langstraat is the center of the footwear industry and that Twente, Tilburg and Helmond are synonymous with the textile industry, this image actually reflects the situation prevailing around the turn of the century. In 1906, 80 percent of the footwear production was in the province of Noord-Brabant, up from only 50 percent in 1858. Shoemaking and leather-working were still found as small-scale crafts throughout large parts of the South and the North at that time. This was also the case in the textile industry (Everwijn 1912, Fig. 2.3).

The appearance of the factory implied a functional as well as a spatial concentration. The large corporation took the lead in the textile industry and thus ushered in the rise of new towns which attracted migrants from the rural areas. This process of functional and spatial concentration (outside the West) brought an end to rural crafts and made the rural economy agrarian by default, as agriculture was the only source of livelihood left. In Holland, the westernmost part of the Netherlands, the development of the crafts industry had been retarded in rural areas in favor of the cities.

The first attempts at industrialization outside the West, as in the relatively large textile mills in Twente and central Noord-Brabant in the 1830s and 1840s, rallied a renaissance in the peripheral states (Keuning 1955) after having been dominated in the crafts sector by the cities in Holland during the Golden Age. Yet when the ports developed around 1870 and the western cities attracted a stream of migrants, another spurt of spatial concentration took place between 1870 and 1910. Clear evidence of this concentration process is provided by the growth of on-site processing of

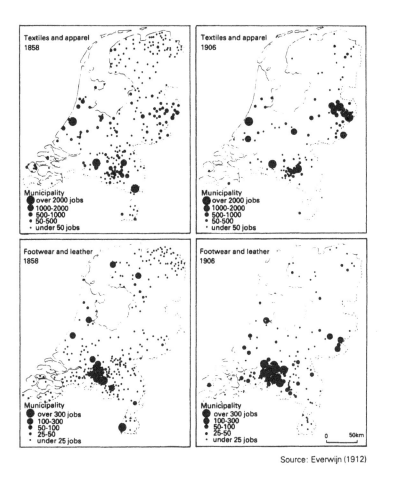

Source: Everwijn (1912)

Fig. 2.3   Locational pattern of the textile industry and
the footwear and leather goods sector, 1858 and
1906

raw materials at dockside, the location of shipyards and
chemical industry adjacent to the ports, and the firm place
in the metropolitan labor market taken by the apparel and
the printing industries as well as the 'basic' economic
services (transport, banking and insurance, etc.). A widely
branching railway network and the presence of a large reser-
voir of labor allowed the eastern and southern peripheral
areas to be drawn into the industrialization and concomitant
urbanization processes. And indeed, after 1910 a trend
toward deconcentration set in (Van der Knaap 1978).

The process of deconcentration engendered a labor-
oriented industrialization drive in the eastern and southern

areas, partly supported by the incipient large industrial firm. In the North and in Zeeland industrialization was clearly retarded. Yet there too, small-scale crafts were snuffed out. In absolute terms, parts of the North and Zeeland actually grew. In the long run, the cities comprising De Randstad and those in peripheral Noord-Brabant achieved a higher position in the urban hierarchy. This was also the case for the large cities in Gelderland and the new industrial cities of Twente and the southern coalmining areas. Cities in the delta and the riverine areas, in the Green Heart of De Randstad, in the North, and in De Achterhoek (eastern Gelderland) all moved down this hierarchy (Deurloo & Hoekveld 1981: 66). Table 2.2, however, shows that the degree of deconcentration should not be exaggerated. The West nearly monopolized the 'basic' business services of national renown. A study by Steigenga (1958) revealed that after 1930, the West exchanged its labor-intensive, traditional manufacturing sectors for rapidly growing industrial port activities.

The rise of the large corporation apparently did not follow a consistent spatial pattern. At least three types may be differentiated:

a. Corporations that sought to tap the labor pool found in the eastern and southern regions, where the sandy soils could not support a growing agricultural population, and that were dedicated to mass production of commodities for the national market. Later these corporations showed an international orientation, even establishing branch plants abroad. An example in the electro-technical field is Philips (founded in 1893); ENKA (established in 1913 and now part of AKZO) exemplifies the rayon fiber industry (Sterkenburg 1938, Stulemeier 1938). They created their own production environments

Table 2.2 Employment in manufacturing in 1899 and 1930, by region (absolutes and percentages).

	1889 abs.	1930 abs.	1899 perc.	1930 perc.	1930 (1899=100)
North	71,751	108,446	12.0	9.3	151
East	113,548	223,648	19.0	20.2	197
Eest	298,225	572,591	49.9	49.2	192
South/Southwest	113,993	249,476	19.1	21.3	219
The Netherlands	597,517	1.154,161	100.0	100.0	193

Source: De Vries 1977, 22

in Eindhoven and Arnhem, respectively. Philips expanded in and around Eindhoven and by 1929 already had 27,000 employees; ENKA employed 12,000 by then, distributed among locations in Arnhem, Ede, Breda and Nijmegen. During the Depression, the labor force declined due to rationalization (at ENKA) and/or transfer of production capacity to foreign locations (Philips).

b. Corporations that sought out good transport locations, such as shipbuilding and the associated machinery assembly plants, by far the preeminent capital goods industry in the Netherlands. Gradually a pool of skilled labor was formed in a number of areas, particularly in the ports of Rotterdam and along the adjacent waterways of the Noord and Merwede Rivers. This availability of trained personnel gained in importance alongside the key factor of accessibility. A secondary factor was the presence of supporting industry, namely suppliers of metals. This caused shipyards and machine shops to grow side by side. A case in point is the establishment of various firms that later amalgamated into the Unilever corporation; the port of Rotterdam was their common ground for existence.

c. Resource-based enterprises. Some of these enterprises have their head office in the Netherlands but their main activities in the Dutch East Indies or worldwide, such as the Royal Dutch Shell petroleum group (Gerretson 1937-1942) or the Billiton tin corporation, which is now part of Shell. The oil-refining and petrochemical industry in the Netherlands only became important after the Second World War. The Netherlands proved to be a country without sufficient mineral deposits to warrant founding a large corporation to exploit them, with the exception of modest veins of coal in Zuid-Limburg. Foreign companies started to mine coal there in the 1890s; after 1901 the Dutch State Mining Company (DSM) was also involved, as Dutch entrepreneurs showed no interest (Raedts 1974). The State also participated in other heavy industries. It acquired a minority interest in Hoogovens, the Dutch iron and steel producer founded in 1918 in IJmuiden, in order to stimulate private parties to take initiatives (De Vries 1968). However, neither Hoogovens nor Staatsmijnen became the core of an integrated industrial complex, as found for example in Germany within the 'Konzernen'. Extraction of salt (at Boekelo in Twente) and oil (in southeast Drenthe) was initiated, respectively, just before and during the Second World War.

2.3 Harmonious Industrial Emancipation, 1952-1963

With the admirable feat of achieving a successful first phase of industrialization (1949-1952) behind them, both government and the business community had more confidence in the future. Secretary of Economic Affairs Zijlstra introduced four major aspects of the social-economic policy in Parliament in 1952: 1. maintenance and improvement of the export position; 2. increase in productivity; 3. an effective industrialization policy; and 4. support for the goal of European unity (Van der Linden 1985: 143).

The Netherlands moved toward conformity with the West European pattern, and the emancipation from its parochial past was complete around 1963. By that time the Netherlands had built up its heavy manufacturing sector and a strong export position. The upward slope of the curve depicted in Fig. 2.2 is clear evidence of the rising production volume in the period 1929-1963. By the end of that period, the wage level had reached West European heights and the time of 'quiet' on the labor front had come to an end. The Netherlands had joined the ranks of the West European industrial nations.

The period 1952-1957, which Wijers (1982) called a second stage of industrialization, has a highly quantitative thrust. Priority was given to the creation of as much employment as possible for a population that was growing rapidly, by West European standards. Dutch society still showed adequate support for wage and price controls (including rents). The ever-present demand for social security began to bear fruit under successive cabinets of Prime Minister Drees, in which the religiously aligned parties shared power with the socialists. Construction on the edifice of the Welfare State was progressing steadily.

The period 1957-1963, the third stage of industrialization, was expected to mark the beginning of a qualitative change. Experts were already well aware of structural weaknesses in the Dutch manufacturing sector (Kohnstamm 1947), but attempts to strengthen it proved ineffective. The gaps, such as the absence of heavy industry and a limited range of high-quality products due to a lack of trained research personnel, required an industrial policy targeted to specific industrial sectors and possible State participation. A successful integration of the Netherlands into the European Economic Community (Treaty of Rome 1957) would only be feasible after the emancipation of its economy was complete.

Industrialization Policy

As the labor force was expected to grow, the fourth memorandum on industrialization (1953) set the goal to create 175,000 jobs, of which 105,000 were in manufacturing, during the period 1952-1957. The sixth industrialization memorandum (1958) revealed that this goal had been surpassed. Unemployment declined to nearly one percent in 1956, a record low. In fact, a number of labor submarkets had become very tight.

Therefore the fifth industrialization memorandum (1956) tackled the problem of the quality of the industrial expansion; a more job-oriented training policy and a more intensive research effort were called for. Yet the ensuing policy was limited to creation of the preconditions for industrialization, including the expansion of the central organization for applied research (TNO), and the extension of industrial guarantee and development loans and fiscal measures as investment incentives. On the other hand, the government shied away from implementing selective, customized government intervention programs. Instead, priority was given to fostering a healthy industrial climate. There were some exceptions, however, whereby governmental inputs in projects such as NV Breedband in IJmuiden (a modern steel rolling mill, subsequently sold to Hoogovens) and Royal Dutch Soda Industry in Delfzijl were made to bolster heavy industry. The Industrial Licensing Act was applied in eleven cases (Van den Hoek 1956). This act permitted government to curtail the freedom of firms to locate and expand industrial establishments on the grounds of the industrialization policy (i.e. to protect industrial sectors new to the Netherlands) and the fair competition principle (in the face of impending overcapacity).

The situation compelled a transition from a quantitative industrialization policy to one more concerned with quality: mechanization (substitution of the factors of production) and a qualitatively changing and growing labor supply necessitated the adjustment of policy directions. It is curious that the differential growth observed among the industrial subsectors did not affect policy during the third stage of industrialization (1957-1963). In this context, Wijers (1982: 39) made an interesting observation: in the third stage of industrialization the sectoral dimension continually recedes and the attention is divided between the macro and the regional levels, whereby the latter tends to win ground (in the form of dispersal policy). This is remarkable, in light of our observation of a diversity in the dynamics manifested among industrial subsectors and the retardation or stagnation of the growth in some industries. It later proved that the government had become aware of the need for restructuring too late.

When the policy aims are confronted with their outcomes during three periods of industrialization (Table 2.3), it appears that the goals have been reached in almost every case, and sometimes even surpassed, as between 1952 and 1962. However, the high expectations for job creation in the third stage of industrialization were not met, at least not for manufacturing. The tertiary sector, however, compensated for this shortfall. The structural changes in the economy are illustrated by the fact that during the third stage of industrialization, half of the rapidly growing labor force was funneled toward the expansion of services in the private sector.

Table 2.3 Realization of goals in three periods of industrialization.

	1 Jan. '48 to 1 July '52		1 July '52 to 1 July '57		1 July '57 to 1 July '62	
	Goal	Reali- zation	Goal	Reali- zation	Goal	Reali- zation
jobs (x 1000 man-years)	125	135	105	208	140	127
job productivity (percentage increase)	23	23	15	23	12	18
gross manufacturing product (percentage increase)	34	35	22	40	21	27
gross investment in manufacturing (x billion guilders)	5.8	5.5	8.5	9.3	11	12.7

Source: Achtste Industrialisatienota 1963

Social Context

Because socioeconomic policy was so strongly focused on the phenomenon of industrialization, the influence of the social context that made industrialization imperative and set the conditions for its realization tends to be underestimated. The equilibrium achieved in the balance of payments was due to the increasing efficiency in agriculture and the traditionally capable private service sector, both of which were able to profit from the economic recovery in Western Europe.

The unification of Western Europe was vital to the Netherlands. The cooperation with Belgium and Luxembourg in the Benelux (1944) was but a modest first step. Soon it became apparent that the Organization for European Economic Cooperation formed a precondition for participation in the European Recovery Program. The European Coal and Steel Association, established in 1951, did not initially appear important for the Netherlands, although this was to change in the 1960s. But when the same member nations formed the European Economic Community in 1957, its potential importance was immediately apparent. To illustrate the role of the

Netherlands in the process of European unification, it may be pointed out that the seeds of European agricultural policy were sown largely by the Dutch.

When the Netherlands became embedded in international cooperative structures, compliance became more mandatory than previous forms of participation. Cooperation reduced the maneuverability of the Dutch, but at the same time it provided a large and growing communal market to compensate for the limitations of the domestic market. Despite the advantages that came along with the open economy, the Netherlands had to accept painful restructuring as part of the package.

In addition to the new international challenge, there was still a domestic problem to be resolved: how to provide work and social security for a rapidly growing labor force? The question also arose whether, besides its advantages, industrialization might entail some negative effects as well. In spite of the urgency of this question, the alternative to industrialization, emigration, began to lose hold after 1952 (Heeren 1985).

Moreover, after 1955, employers' organizations opposed the promotion of emigration in light of the increasing tightness of the labor market. Nor did the population policy promise relief. This policy had been synonymous with emigration policy for a long time, because there was no political support for promoting a reduction in the birthrate. Arguments in the defence of a decline in population were only voiced in government in the mid-1960s. Fortunately a hard confrontation with a situation of overpopulation, whereby, according to current standards, the subsistence resources would not be capable of supporting the population, was preempted by the sucessful industrialization and the unanticipated strong growth of the service sector (including government) (Table 2. 4). Redundancy due to the mechanization and rationalization of agriculture was absorbed by the labor market in most regions. It is remarkable how steadily manufacturing and services grew in each of the successive five-year periods between 1950 and 1965; the conjunctural ripples had not yet disappeared completely, but they had been brought reasonably under control. The importance of industrialization differed considerably by region (Fig. 2. 4). The transformation from agriculture to industry was most striking in the North; in the West, industry grew in terms of the number of jobs, but the economy remained predominantly service-oriented. The South remains the most industrialized part of the country and, in absolute terms, this role has been reinforced. The East is less industrialized.

The Secretary of Economic Affairs Van den Brink attacked the 'white-collar mentality' in the first industrializa-

25

Table 2.4   Employment by sector, 1950-1965.*

Sector	1950 abs.	1950 perc.	1955 abs.	1955 perc.	1960 abs.	1960 perc.	1965 abs.	1965 perc.	indices (1950 = 100) 1955	indices (1950 = 100) 1960	indices (1950 = 100) 1965
Agriculture	582	16	532	14	465	11	388	9	91	80	67
Manu-facturing**	1495	41	1630	42	1715	43	1886	43	109	115	126
Services	1304	36	1385	36	1512	38	1711	39	106	116	131
Government	265	7	300	8	326	8	361	8	113	123	137
Total	3646	100	3847	100	4018	100	4346	100	106	110	119

*   labor volume (man-years) x 1000
** mineral resources and public utilities included

Source: Centraal Planbureau, De Nederlandse economie in 1973

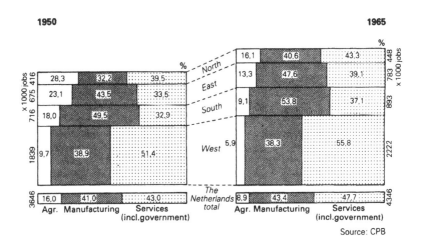

Fig. 2.4   Regional employment per sector, 1950 and 1965

tion memorandum (1949). Educationalists opposed vocational schooling that would only be functional in manufacturing; all-round education should be expanded to meet the demands made by a changing society (De Liagre Böhl et al. 1981). An important obstacle, however, was the lack of adequately trained technical personnel. A significant effort was made to rectify this situation: the number of lower vocational schools (LTS) doubled in the 1950s; the number of certificates issued upon completion of apprenticeship rose from 2000 in 1948 to 15,000 in 1961; and a new type of mid-level vocational school (UTS, now MTS) was established, with 31 schools in 1961. Although the number of higher-level polytechnics (HTS) grew, the eighth memorandum on industrialization (Achtste Industrialisatienota 1963: 50) nonetheless noted a shortfall. The Technical Universities (TU) had to wait until the third stage of industrialization to undergo structural expansion (the establishment of TU Eindhoven first, then TU Twente). Partly for this reason, industrial research was underdeveloped. Only the large corporations invested in research.

2.4 Evaluation and Summary

In a period when it is being debated which path the Netherlands should take as an industrial nation, it is worth looking back four decades in time to the years of postwar recovery and reconstruction. A remarkably strong growth in population, an unfavorable balance of payments and redundancy of labor in agriculture made an industrialization offensive imperative. The goals set in the industrialization policy were amply reached.

The Netherlands is an industrial late bloomer. It was only during the upswing of the third Kondratieff cycle in the final decade of the last century that the industrial breakthrough was realized. The postwar industrialization policy was to profit from the upswing of the fourth Kondratieff cycle. In general terms, this trend of the long waves applies to the Netherlands. Both upswings were characterized by basic innovations by large Dutch firms (Philips, AKZO, etc.) or by foreign firms that had been attracted to locate in the Netherlands (for example, foreign firms participating in the port industrialization from the 1950s on). The participation of a nation in the upswing of a new long wave depends as much on the organizational form and the locational choice of corporations as on the introduction of new products.

The theories of Hirschmann and Rostow, which are concerned with industrial breakthroughs engineered by strategic sectors, do not prove to be very relevant to the Netherlands. As far as its industry is concerned, the Netherlands came into the picture too late for an industrialization process based on iron and steel manufacturing, which characterized the second Kondratieff. The 'new' industries at the turn of the century were functionally and geographically less concentrated and less strongly intertwined. Moreover, the Netherlands played a central role in the exchange of industrial products among neighboring countries, which gave the tertiary sector a strong impulse.

The infrastructure, serving both domestic and international linkages, was the ideal catalyst for an industrial breakthrough. The fragmented industrial pattern of the Netherlands was the object of thorough scrutiny in the early postwar industrial policy. In the 1960s, the metallurgical and petrochemical heavy industries matured and the equally necessary improvement in industrial standards began to take shape.

Geographically, the period of the port development and the industrial breakthrough (1870-1913) manifested an increasing spatial concentration of manufacturing, and thus of the population as well. In the growing urban system the 'big three' took the lead and new industrial cities emerged in the East and the South. From the 1920s, even in the face of ongoing urbanization, deconcentration of industry, and concomitantly of population, accompanied the application of electric power (a product of the electrification of rural areas) and the upgrading of transportation (the railroad). In this context, the main thrust of the theories of Myrdal and Pred applies to the Netherlands during the third Kondratieff. Indeed, both the North and Zeeland still bear the brunt of the backwash effects. Still, at the scale of the Netherlands, the core-periphery dichotomy is much less explicit than in France, for example. The post-war industrialization process would nonetheless prove these theories wrong, at least in quantitative terms. Qualitatively the differences would remain clearly visible.

3 The International Perspective of Dutch Industry

3.1 Industrial Development and Renewal

The concept of 'innovation' plays a key role in discussions about economic development. The official Dutch government memorandum on innovation (1979) defines it as the development and successful introduction of new or improved goods, services, production processes or distribution procedures. The significance of innovation, and of renewal in general, is convincingly formulated in the policy review of technology 1984-1985 as follows: it is generally accepted that the development and application of new technologies and the new products and services they generate form essential preconditions for enlarging that portion of the total package of products of this country which is currently at the beginning of the product life cycle. Like the use of advanced methods of production, such a package is not only increasingly required to retain and to buttress the competitive capacity of business, but it also provides favorable employment and productivity effects on a macro-economic scale.

<u>The Product Life Cycle</u>

The product life cycle (Fig. 3.1) assumes that a product, narrowly defined (a particular car or light bulb), passes through four stages during its existence. In the introduction or innovation stage it is brought onto the market. It usually meets with a passive reception by potential buyers. If the product is a substitute for another item, it will encounter resistance on the part of the producers involved. The difficult start is partially due to the fact that the product has some wrinkles that still have to be ironed out. After surmounting the initial difficulties, it can get a foothold, especially when quality and price are in its favor. The product then enters the expansion stage. Sales pick up rapidly. Technically it is almost perfected, which implies that technical aspects of production become less important to the manufacturer and the importance of market-

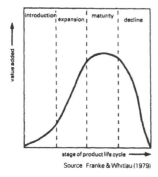

Source Franke & Whitlau (1979)

Fig. 3.1   The product life cycle

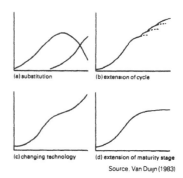

(a) substitution

(b) extension of cycle

(c) changing technology

(d) extension of maturity stage

Source. Van Duijn (1983)

Fig. 3.2   The product life cycle:
some variations

ing and distribution aspects increases. The growing market
can induce the enterprise that launched the product to
cultivate markets elsewhere and eventually to export it.

On the other hand, a growing sales record attracts
competition. As long as the market is expanding, this is no
great problem. It is a different matter when the market
begins to show signs of saturation, whereby sales first grow
at a slower rate, then stagnate and subsequently decline.
In the stage of maturity, this saturation process takes
hold. Through increased competition the profit margins
decrease. In reaction, attempts are made to hold the costs
down, for example by streamlining the production process
which by this time has become standardized. The end of the
life cycle is the stage of recession, in which sales decline
and demand is limited to replacement purchases. If the
product is to survive this stage, then an optimal standard-
ization and rationalization are imperative. Even then,
production proves to be unprofitable in many places unless
protectionistic measures are taken. Production concentrates
where the costs are lowest.

The product life cycle can be expressed in spatial
terms. It is assumed that the new product is preferably
introduced in large, affluent markets, hence in the economically most developed countries or parts thereof, where the
innovation has often taken place. Contact with and proximity
to the client is important in this stage to perfect the
product. In the expansion stage the production is greatly
expanded in the country or areas involved, whereby an increasing proportion is sold elsewhere. In this stage the first
competitors intrude. In the stage of maturity it becomes
increasingly difficult for the original producer to stave
off competition, particularly when it produces at a location
where costs are significantly lower. When the production
process is standardized, production in the economic core
areas offers few advantages; instead, it poses disadvanta-

ges, due to the higher costs of production. Some producers choose to establish branches in peripheral areas of their country where the costs are lower. This can lead to a spatial division of the functions of the enterprise. Such a move can be attractive for production in particular. The headquarters with the central management and departments such as research and marketing mostly stay at the original location. The trend to seek favorable, cheap locations for production activities continues even more fervently in the stage of recession. Eventually, it may happen that all production units will be closed in the area where the manufacturing of the product in question began. This fate may also lie in store at a later stage for the production units in the low-cost areas when they cannot keep up with the competition of even cheaper locations. In the US, for example, this can cause an initial shift of production to low-cost areas within the country and subsequent moves to low-cost nations in Central America and then on to even cheaper locations in Southeast Asia. Eventually the demand in the original production area, the economic core region of the US, will be met by imports.

The sequence of events in the product life cycle is, of course, generalized; many products take a different course. Some of the possible variations are depicted in Fig. 3.2. But despite these variations and possible criticism (Kamann 1985, Taylor 1987), the notion of life cycle may be related to the rise and fall of branches of industry and to the competitive position of Dutch industry in the international arena.

Table 3.1 Labor productivity and manufacturing employment EC (annual percentage growth).

	1960-68	1968-73	1973-79	1979-84
Productivity				
EC	4.4	4.4	2.3	1.5
Japan	8.9	7.7	2.9	2.6
US	2.6	1.3	0.2	0.7
Jobs in manufacturing				
EC	0.3	0.3	-1.0	-2.3
Japan	4.0	2.7	-0.4	1.4
US	1.9	1.0	1.5	-1.2

Source: Albeda in Economisch Weekblad 22 February, 1986

31

3.2 Industry in the Netherlands: Strong and Weak Points

<u>The International Picture</u>

In the period 1960-1985 there were shifts in the position of the three most important poles in the international economic arena. Japan recorded the strongest increase in productivity and industrial employment. The EC showed a satisfactory development in productivity. In comparison with the other blocs, though, the EC saw its industrial employment stagnate (Table 3.1). Partly for this reason, the unemployment rate in the US was clearly lower than in the EC in 1985 (7.25 percent against 11.25 percent). The fact that Western Europe has trouble maintaining its position is also demonstrated by the list of the most important producers of silicon chips (Table 3.2), a product that is still in the growth stage of

Table 3.2   Top-10 chip manufacturers 1986.

Firms	estimated sales (x billion)
1. NEC-Japan	2,865
2. Hitachi-Japan	2,270
3. Toshiba-Japan	1,975
4. Texas instruments-US	1,970
5. Motorola-US	1,820
6. Fujitsu-Japan	1,560
7. Philips*-The Netherlands	1,100
8. Intel-US	1,005
9. Matsushita-Japan	985
10. National-US	970

* Inc. Signetics-US

Source: IC Engin. Corp.

its life cycle. Japanese enterprises are in the lead. Of the European producers, only Philips (including the US subsidiary Signetics) is included among the top ten. A more detailed overview is given in Table 3.3. For a number of countries, two criteria are used to determine a trade achievement score: the export specialization coefficient and the relation between export and competing import per commodity group. The classification by commodity group is derived from the OECD and is based on (somewhat dated) statistics on the intensity of R&D (research and development) expenditure in the US. On the whole, Japan comes out on top. In Western Europe, the Federal Republic of Germany, Sweden and France score high. The Netherlands takes a very modest position and the score in the high-value sector is outright poor. Only in

Table 3.3 Trade achievement score[1].

| Level of technology | achievement level | | | | | | | | achievement progress | | | | | | | |
| | Western Europe | | | | | | | | Western-Europe | | | | | | | |
	US	JAP	FRG	FRA	UK	SWE	BLEU	NETH	US	JAP	FRG	FRA	UK	SWE	BLEU	NETH
High value																
aerospace	++	--		+	++	--	--	--	--		+	+	+	+	+	-
office machines and computers	++		-	-		-	--			+	+	-	+	-	=	=
electronic components		++			-		-			+	-	=	-	-	-	=
pharmaceuticals and drugs	+	--	+	++	+	+	+		-		-	+	-	+	+	-
instruments		++	+	-	+	--	--			-	-	-	-	+	+	-
electrical machines	+	++	+	+	+	-	-		-	+	-	=	-	=	=	-
Intermediate value																
automotive	-	++	++	+	-	+		--	-	+	+	-	-	+	+	+
chemicals	+	+	+		+	--	+	++	=	-	-	+	+	=	=	+
other manufactures	-		-			--	-	--	+	+			-	+	-	=
non-electrical machinery	+	+	++		+	+		--	-	+	-	-	=	+		+
rubber, plastics	--	+	+			--	+	-	-				=	+		-
non-ferrous metals	-	--	-	+	+	+	+		+	+	+	+	=	+	-	+
Low value																
stone, clay, glass	-			+		-	+	--	=	-	-	+	-	+	-	-
shipbuilding		++	+	+		++	-		+	-	-	+	-	-	-	-
ferrous metals	--	++	+	+	-	+	++		-	-	+	+	=	+	-	-
metal products			+		+	+			=	-	-	=	-	+	+	
paper and printing			-		-	++			=	-	+	=	+	-	=	-
wood, cork, furniture	--	--	-		--	++		--	=	-		=	=	-	-	=
apparel, footwear, leather goods		--	-				+	-	-	-		-	-	=	-	-

1 The trade achievement score is determined by the export specialization coefficient and the relation between export and import value. No sign is shown if the two measures give contradictory results. Data for the achievement level refer to around 1982; the progress is based on data for a few years prior to 1982.

Source: C.P.B. (CEP 1986, 280)

Table 3.4 Production and labor productivity, 1960-1985, the
          Netherlands and EC (annual percentage fluctuation).

	production	labor productivity
1960-1973		
The Netherlands	5.0	3.8
EC	4.6	4.4
1973-1981		
The Netherlands	1.9	1.2
EC	2.0	1.9
1981-1985		
The Netherlands	0.7	1.0
EC	1.6	2.0

Source: Bosch et al. 1985, 8

the intermediate category does the Netherlands take a relatively more favorable position. Specifically, the internationally strong position of the Dutch chemical industry shows up there.

Just as the EC is struggling to keep up in the international context, the Netherlands seems to have trouble maintaining its position within the EC. In the period 1960-1973, the growth of production in the Netherlands was even higher than in the EC as a whole. The growth in labor productivity was about equal to the EC level. Afterwards, the degree to which the Netherlands lagged behind the EC increased continually (Table 3.4).

There are some major drawbacks to comparing the Netherlands with the EC, the US or Japan. Countries with a small domestic market often generate their economic growth through export. Besides a high level of export, those countries are also big importers. As a result, the indirect effects of export (the multiplier effects) are often limited because these leak away to foreign countries through import. On the other hand, it is instructive to compare the development of the Dutch economy with the economies of other small European nations.

In the period 1967-1973, the Netherlands showed a satisfactory development of the GNP relative to other small European countries. Only Austria attained a higher level of growth. This was followed by a downturn in all countries. The reversal was more acute in the Netherlands than elsewhere, especially in the period 1979-1983. This resulted in an extremely high level of unemployment, although it should be noted that international unemployment figures are not

Table 3.5 Comparison of the Dutch economy with the economies of other small EC members and the EC as a whole.

	Belgium	Denmark	Austria	Switzerland	EC	The Netherlands
Gross Domestic Product (GDP)						
- annual growth						
1967-1973	5.2	4.0	5.7	4.3	4.9	5.5
1973-1979	2.4	2.1	2.9	-0.4	2.5	2.5
1979-1983	0.9	1.2	1.5	1.1	0.6	-0.1
Gross investment						
- annual growth						
1973-1979	1.4	-1.2	0.9	-2.9	0.5	0.0
1979-1983	-4.4	-5.4	-1.3	2.8	-1.2	-3.7
Export						
- goods and services as % GDP	73.2	35.6	41.4	35.4	29.9	57.5
- export structure for commodities (%)						
agricultural products	10.3	31.9	4.3	3.3	11.1	20.0
raw materials	3.2	7.5	7.6	1.6	3.3	5.7
energy	9.1	3.3	1.8	0.1	9.6	24.2
chemical products	11.7	7.5	9.2	20.6	11.5	15.2
machinery and vehicles	21.9	25.0	27.4	32.2	32.9	16.1
other industrial products	43.9	24.8	49.7	42.1	31.5	18.8
Employment (%)						
- agriculture	3.0	7.3	8.8	7.0	7.9	5.0
- manufacturing	33.4	29.3	39.5	39.3	37.1	30.2
- government	18.0	29.6	18.7	10.2	17.4	15.4
- other services	45.6	33.8	33.0	43.5	37.6	49.4

Source: Centraal Planbureau, Centraal Economisch Plan 1984

entirely comparable due to differences in definitions (Table 3.5).

Table 3.6 presents a more recent picture of the position of the Netherlands in Western Europe in terms of some other indicators. The picture includes a moderate degree of economic growth and a rise in industrial production along with a low inflation and a sizeable balance of payments surplus.

A New Industrial Impetus

The disappointing development of Dutch industry did not go unnoticed. The Scientific Council for Governmental Policy (WRR), the government's think tank, played a key role in evaluating it; in 1980 this agency reported its deliberations on the role and future of Dutch industry. The relatively weak position of manufacturing at that time was illustrated in the WRR report by data on the increasing penetration of the Dutch market by foreign manufacturing firms. The WRR also found a weakening in the export position of Dutch manufacturing, particularly in comparison with other industrial nations.

This weak position of manufacturing in the Netherlands is related to five factors:

- the steady rise in wages during the period 1960-1980, which led to a higher cost of labor per unit of production as well as to a steep rise in productivity, via substitution of capital for labor;
- the composition of the Dutch export package. Agricultural and energy (natural gas) products are well represented in this package. The export of capital goods, on the other hand, is poorly developed ('other manufacturing' in Table 3.5).
- the real increase in value of the Dutch guilder since 1971. This brought a relative rise in the price of Dutch products, which hindered export and eased import, since the foreign products became less expensive in Dutch guilders;
- the one-sided geographical orientation of Dutch export toward the EC member states. Not only the most important export sectors of agriculture and energy, but also manufactures flow in high volume to the EC; in contrast, the volume of export to the areas of economic growth on the Pacific rim is low.
- the institutional stagnation of the Dutch economy. The WRR sees this as the cause of the inadequate functioning of the labor market, the reason why individual firms that are ineligible for support do receive aid, the explanation for the absence of the spirit of enterprise and the source of red tape which retards or even

Table 3.6 Economic indicators 1985, European countries.

Unemployment (perc.)		Economic growth* (perc.)		Inflation* (perc.)		Industrial production* (perc.)		Balance of payments (US$ per capita)	
1.0	Switzerland	4.4	Norway	2.2	FRG	5.6	Switzerland	+725	Norway
2.5	Norway	3.2	Switzerland	2.2	The Netherlands	5.4	FRG	+699	Switzerland
2.8	Sweden	3.0	UK			4.8	UK	+409	The Netherlands
4.9	Austria	2.9	Austria	3.2	Austria	4.3	Austria		
6.3	Finland	2.8	Finland	3.4	Switzerland	4.0	Denmark	+212	Germany
9.0	Denmark	2.7	Denmark	4.7	Denmark	3.5	Norway	+ 77	Spain
9.3	FRG	2.4	FRG	4.9	Belgium	3.2	Finland	+ 67	UK
9.3	France	2.3	Italy	5.6	Norway			+ 10	Belgium
10.6	Italy			5.8	France	3.1	The Netherlands	+ 5	France
13.1	UK	2.1	The Netherlands	5.9	Finland			- 13	Austria
15.6	The Netherlands	2.1	Sweden	6.1	UK	2.3	Belgium	- 72	Italy
18.6	Belgium	1.7	Spain	7.3	Sweden	2.0	Sweden	- 96	Sweden
22.3	Spain	1.5	Belgium	8.8	Spain	1.8	Spain	-123	Finland
		1.3	France	9.2	Italy	1.7	Italy	-509	Denmark
						0.3	France		
11.0	Europe (mean)	2.3	Europe (mean)	6.7	Europe (mean)	3.3	Europe (mean)	+ 54	Europe (mean)

* Compared with 1984

Source: OESO

obstructs all kinds of initiatives.

According to the WRR, the goal should be a more diverse industrial structure, a range of higher-quality products (demanding more research and training) and a stronger emphasis on product differentiation and non-price aspects (reliability, service, etc.). This recommendation takes the altered international relation among competitors into account. The emergence of the Newly Industrializing Countries (NICs) and the OPEC member states has cast a shadow on the perspectives for a number of sectors that played an important role in the Netherlands' economic recovery after 1945. These include the competition with the NICs for labor-intensive manufactures (apparel, footwear) and with the OPEC countries for oil and petrochemical products. Consequently, the Netherlands has no choice but to concentrate on products in the early stages of their life cycle.

Reorientation

In response to the report by the WRR on the status of Dutch industry, certain activities were identified as high-potential opportunities. A first inventory was compiled by the Commission for the Advancement of Industrial Policy, also known as the Wagner Commission, named after its chairman (the former president of the board of directors of Royal Dutch Shell). The best opportunities were recognized in products with a growing international sales record (in the first stages in their life cycle) on which the Netherlands had an edge expressed, for example, in a high export/import ratio and by the fact that the Netherlands is one of the world's most important exporters. The industrial activities considered to be good opportunities (Table 3.7 'Wagner list') subsequently formed the basis for the policy of primary concentration areas, an instrument through which the government attempts to improve the structure of the Netherlands' economy.

Geldens (in Franke & Whitlau 1979), a consultant with McKinsey, the well-known efficiency experts, approached the problem in terms of the life cycle. Although this concept refers to individual products, he applies it to industry groups. The categories of introduction, expansion, maturity and decline embody certain criteria (Fig. 3.3). The industry groups can be classified and the economic profile of the Netherlands can be drawn on the basis of these indicators. To buttress the economy, according to Geldens, the government should promote and support activities which are in the early stages of their life cycle. This idea closely resembles the findings of the Wagner Commission.

Table 3.7 List of high-potential activities identified by the 'Wagner Commission'.

Target areas	Activities
- Port areas	. traffic control systems
	. container-handling equipment
	. pipe line transport
	. specialized services
	. consultancy
- Agricultural and food industries	. strain improvement and quality products
	. distribution systems
	. farm machinery
	. know-how horticulture and auction systems
	. pre-fab module factories
- Civil engineering, flood control, water management and continental shelf exploration	. delta technology
	. waterways
	. specialized shipbuilding
	. purification of drinking water
	. treatment of waste-water
	. offshore industries
- Amsterdam Airport	. aircraft industry suppliers
	. airport/freight handling equipment
	. airtraffic control
- Chemical refining	. products of high added value (fine chemicals)
	. processing equipment
	. pre-fab modular construction of chemical factories
	. selective processes (bio-technology)
- Electronics and informatics	. telecommunication systems and hardware
	. robotics
	. teletext, viditel
- Automated business services	. computer-aided design
	. software industry
	. consultancy/systems engineering
	. turnkey systems
- Maintenance and renovation	. maintenance equipment
	. building renovation
	. road and waterway maintenance
- Energy installation and equipment	. heating plants
	. energy-saving techniques and devices
	. coal transhipment and transport
	. natural gas pipelines
	. maintenance of gas turbines
- Medical technology	. measuring and monitoring equipment
	. aids for the handicapped
- Waste processing, recycling and environmental pollution control	. iron and non-ferrous metals
	. separation techniques for domestic and farm refuse

– Defence equipment	. shipbuilding
	. vehicles
	. radar, anti-aircraft apparatus
	. optical instruments
– Construction	. inexpensive home-building
	. wood-frame construction
	. home furnishings and fabrics
	. household energy saving
– Office systems	. reproduction techniques
	. graphic design
	. office automation

Source: 'Een nieuw industrieel elan' 1981, 37/38

Recent Developments

There have been some changes in recent years, due in part to implementation of elements of the WRR report and the recommendations of the Wagner Commission. Although the image of high wages has not disappeared, the cost of labor per unit of production in manufacturing has decreased since 1980. In the international context, wages in the Netherlands were at a high, but not extremely high, level in 1987, which strengthened the country's competitive position (Fig. 3.4). Indeed, after 1981 foreign firms ceased to penetrate further into the Dutch market. Moreover, in the period 1980-1985 the proportion of Dutch goods in the foreign trade between the EC and the US increased. This naturally benefited industry groups that depend on export (Fig. 3.5) and thus the areas where many exporting firms are located. Of course, the decline in domestic demand that resulted from wage controls had a negative impact on nationally or regionally oriented industry groups.

Besides the goal of improving the competitive position of the Dutch economy through wage controls, government policy has recently given much attention to the role R&D can play in generating a high-quality industrial structure by developing new products and production processes. In particular, this implies better and more commercial utilization of the know-how available at the universities and polytechnics (Gibb 1985). Virtually all of these schools now have so-called transfer points to facilitate the connection between industry and institutions of higher learning. Following the example of other countries, initiatives have been taken to establish science parks (as at the University of Groningen) and industrial-technological centers (as at the Technical University of Twente) which are allied to the respective schools.

The trend toward products requiring higher technology and advanced training is not only evident in the Netherlands but is found in all West European countries. Everywhere the

Criteria	Stage of lifecycle			
	Introduction	Expansion	Maturity	Decline
Technology intensity[1]	● ●	●		
Advertising intensity[2]	●	● ●	● ●	
Export share[3]		● ●	●	
Export growth[4]	●	● ●		
Import share[3]			●	● ●
Import growth[4]			●	● ●
Foreign investment[5]		●	●	●
Increase in investment[6]	●	● ●		
Decrease in investment[6]			●	● ●
Increase in jobs[6]	●	● ●		
Decrease in jobs[6]			●	● ●

●   = relatively high

● ● = relatively very high

1 Number of technicians (with at least college degrees) as percentage of all employees

2 Advertising expenditure as percentage of all liquid assets

3 Export or import as percentage of domestic sales

4 Change in share of export or import during the last three years in comparison to the previous three–year period

5 Foreign interest earned plus tax credit as percentage of fixed assets

6 Change in investment or employment levels during the last three years in comparison to the previous three–year period

Source. Franke & Whitlau (1979)

Fig. 3.3   Criteria used to classify industrial sectors according to stage in the product life cycle

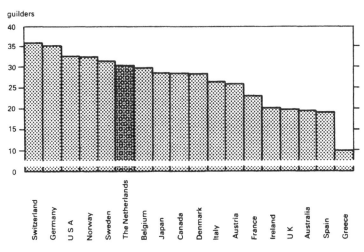

Source: Institut der Deutschen Wirtschaft (IDW)

Fig. 3.4   The cost of labor in manufacturing; the Netherlands in an international perspective

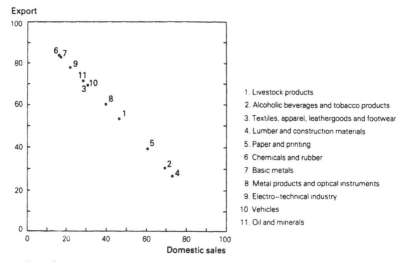

Note: Export includes sale to intermediaries supplying foreign branches of industry, i.e indirect export

Source: CPB 1986

Fig. 3.5   Sales on domestic and foreign markets (as percentage of production value)

outlays for R&D are increasing rapidly. Comparison of Dutch R&D expenditure with that in Belgium, Sweden and Canada reveals that the growth of the R&D budgets of Canadian and Swedish business are clearly larger. In contrast, the contribution of the Dutch government to R&D has grown significantly. The role of government in financing R&D has remained steady in Canada and has even declined in Sweden (Fig. 3.6). It is interesting to note that Dutch R&D expenditures are highly concentrated in five large multinationals: Shell, Unilever, Philips, AKZO and DSM. Of all R&D employment in the Netherlands 80 percent is within these five firms.

3.3 Industrial Structure and Restructuring

The Place of Industry in Economic Development

About 50 years ago Clark and Fischer predicted the transition from an industrial to a service-oriented society. In comparison to the service sector, productivity in agriculture and industry rose steeply, which would subsequently reduce the relative importance of these sectors as employers. The role of services in employment was expected to be buttressed by the rising demand for services that accompanies rising affluence.

Within three decades, the shift in employment has indeed come to correspond closely to the scenario presented by Clark and Fischer (Table 3.8). In the 1950s and 1960s

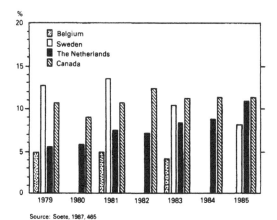

Source: Soete, 1987, 465

Fig. 3.6a R & D expenditure by firms ($ mln, current
          prices)

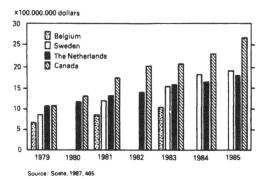

Source: Soete, 1987, 465

Fig. 3.6b Government share in the financing of R & D
           expenditure by firms

agriculture and mining (i. e. the primary sector) were the
big losers. The decline in manufacturing materialized at a
later stage. Especially during the long recession which
started in 1973 many manufacturing jobs were lost. Together
with the other big loser, the construction industry, these
sectors were responsible for a net fall in employment during
the decade 1973-1983 in the Netherlands. The continued shift
toward services, which seemed unquestionable, lost momentum
after 1973. This applies specifically to the (private)
tertiary sector, including trade, transport, banking and
insurance, business services, and less to the quaternary
(social) service sector and its employment. The developments
of the past 10 years demonstrate a stabilization of employment in agriculture (Table 3. 9). The rate of growth in the
service sector continued to taper off. The development in
industry after 1984 is noteworthy. After years of erosion,
industry has been showing persistent growth in employment.

Table 3.8 Employment by sector, unemployment and labor supply, 1953-1983.

	1953/1963	1963/1973	1973/1983
	annual fluctuation (x 1000 man-years)		
Employment			
agriculture	-13,0	-11,1	- 3,6
manufacturing	17,9	- 8,1	-27,6
energy	- 0,5	- 3,3	0,1
construction	9,2	5,0	-13,7
tertiary	27,2	20,3	3,6
quaternary	6,3	16,6	12,3
Total firms	47,1	19,4	-28,9
government	7,6	9,6	12,1
unemployment	- 5,2	11,7	65,0
Labor supply	49,2	40,7	48,2

Source: CPB (CEP 1986)

Table 3.9 Employment by sector 1977-1987 (x 1000 man-years).

	1977	1980	1984	1985	1986	1987
Wage-earners	3364	3468	3192	3246	3324	3358
agriculture	70	71	72	73	74	75
manufacturing*	985	939	819	833	848	856
energy	61	64	66	66	67	67
construction	385	400	285	288	301	304
services	1863	1994	1950	1986	2034	2066
tertiary	1381	1477	1412	1440	1477	1496
quaternary	482	517	538	546	557	560
Self-employed	640	625	606	607	609	610
Government	676	714	730	736	737	736
Total	4680	4807	4528	4589	4670	4704

* Excluding oil refineries

Structural Changes 1953-1973

In reviewing a decade of continual growth, 1953-1963, there is no trace of a uniform pattern (Table 3. 10). Stagnation in the textile industry and coalmining showed up early, warning of the impending large-scale restructuring process. A number of industry groups demonstrated a relatively low level of growth in employment, production (Fig. 3. 7) and, to a large extent, in investment activity. These include the apparel and footwear industries as well as beverage and tobacco products and specific sectors of the transport industry (shipbuilding). The real wage development and investment quote, however, do not in general correspond to trends in employment and production; they do, however, in textiles and beverages and tobacco products. Eventually more in-depth

Table 3.10 Labor, production and investment in industry, 1953-1963 (in percentages).

Industrial sector	employment increase[1]	production growth[2]	labor costs increase (real)[3]	invest ment quote[4]
Food				
animal products	1.6	5.1	11.1	14.6
other products	-0.1	3.2	4.2	10.9
beverages and tobacco	0.3	6.2	6.6	10.5
textiles	-1.0	3.3	6.8	11.9
apparel and footwear	0.3	3.1	3.3	3.4
paper	3.7	8.6	7.6	17.4
chemicals	4.0	9.6	8.7	20.5
oil refining	5.2	10.8	7.1	71.9
metallurgy	4.3	10.2	5.4	27.6
metal products and machinery	2.4	6.6	5.3	8.2
electro-technical	5.4	14.0	8.5	7.6
transport equipment	1.7	3.1	1.9	8.0
other	1.5	6.4	5.7	8.8
total manufacturing	1.7	6.1	5.9	11.8
manufacturing excluding construction	1.5	5.9	5.5	14.8

[1] Wage-earners and self-employed (labor volume).
[2] Value added (1973 price level).
[3] Wage level per employee as fraction of cost of production.
[4] Gross investment as a percentage of value added (factor costs, 1963 prices, buildings excluded).

Source: CPB 1986

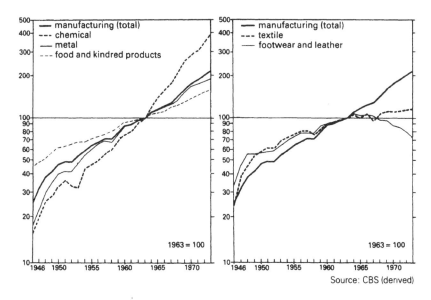

Fig. 3.7 Production indices for the manufacturing sector, 1946-1973; growing and declining sectors

45

investments were made, though. The degree to which this varies may be demonstrated by the so-called growth sectors when these are identified on the basis of growth in jobs and production. Thus defined, the oil refineries, chemical industry and electro-technical industry are prime examples of growers, but their investment quotes differ greatly. A capital-intensive industry like oil refining contrasts sharply with the labor-intensive electro-technical industry. The line of development diverges according to the economic indicator used. The phenomenon of growth thus has various dimensions.

In this era of growth, a split among industry groups emerged. This may be illustrated using employment data for the periods 1953-1958 and 1958-1963. A division by industry groups puts some of the previous findings into perspective (Table 3.11). The growth of jobs in apparel manufacturing is noteworthy. It lasted until 1963 and then deviated from the

Table 3.11 Manufacturing employment 1953-1963: expanding and stagnating industries.

	1953	1958	1963	1958	1963
	abs. (x1000)			(index 1953=100)	
expanding industries					
apparel	55,9	61,7	74,6	110	133
chemicals*	50,1	63,7	77,4	127	154
metallurgy	22,6	25,5	32,7	113	145
machinery	72,0	84,2	103,2	117	143
transport equipment**	21,0	21,8	26,9	104	128
electro-technical	58,6	81,2	102,3	139	175
metal products	43,5	52,7	62,8	121	144
printing	28,8	39,3	46,4	136	161
cement	11,3	12,6	13,9	111	123
total	363,8	442,7	540,2	122	148
stagnating industries					
footwear	16,4	17,9	15,8	109	96
cotton, rayon, linnen	49,0	49,6	47,3	101	96
wool	18,6	18,1	18,4	98	99
bricks	13,3	12,0	11,1	90	83
mineral resources	59,1	61,4	53,2	104	90
shipbuilding	52,1	56,4	50,9	108	98
total	208,5	215,4	196,7	103	94
share of expanding industries	39,5	42,9	46,5		
share of stagnating industries	22,6	20,9	16,9		
total industry (public utilities excluded)	921,8	1030,9	1161,0	112	126

*   Including oil refining
** Excluding shipbuilding

Source: C.B.S.-Alg. Ind. Statistiek

Table 3.12 Employment by sector, 1973-1985.

	1973	1979	1985[1]	1973/1979	1979/1985
	1000 man years			annual perc. growth	
agriculture	309	280	271	-1.6	-0.5
manufacturing	1144	994	860	-2.3	-2.4
energy	66	62	66	-1.0	1.0
construction	465	459	327	-0.2	-5.5
services	2097	2275	2286	1.4	0.1
tertiary	1649	1740	1706	0.9	-0.3
quaternary	448	535	580	3.0	1.4
total firms	4081	4070	3810	0	-1.1
government	612	703	732	2.3	0.7
total	4693	4773	4542	0.3	-0.8

[1] Estimate.

Source: CPB (CEP 1986)

trend followed by other industry groups that had reached the stage of stagnation before or around 1963 and that had already experienced moderate growth in the 1950s (footwear and textiles, mining and shipbuilding). In both periods, the growers, primarily the chemical and electro-technical industries, were unremittent in generating jobs. The printing industry was also one of the branches that experienced a growth in employment of more than 50 percent. Industry on the whole registered 'only' 25 percent.

Summarizing, a picture of differential growth emerges. There were expansive and stagnating industry groups. Their dynamics had already diverged in the third stage of industrialization, when signs of restructuring were apparent. The image of growth is thus not as uniform as it is sometimes portrayed to be. In addition, there are two classes of growers: capital-intensive and labor-intensive industry groups. The question is, how were these developments related to the industrialization policy and to the wider social context?

Developments after 1973

Reviewing the facts, it appears that between 1973 and 1985 industry lost a quarter of its jobs, and construction even lost 30 percent! Agriculture was able to cut its losses (Table 3. 12). Until 1979 the growing tertiary sector compensated for the decline. Despite recent economic recovery, it is clear that in the first half of the 1980s, industry is continuing as ever to shed labor; even the tertiary sector is showing a gross reduction in employment for the first time in history. The total number of jobs had declined significantly since 1979 and only recovered after 1984.

Besides these disappointing trends in employment, the Netherlands also had to absorb a demographic boom for the second time. The first one appeared in the 1950s (Table 3.8). In the sixties, the number of new jobs eventually surpassed the growing labor supply. In the seventies, however, no further increase in employment compensated for the vigorous growth in the supply of labor (Table 3.12). On the contrary, unemployment rose considerably at that time. It is useful to call attention to a qualitative change in the composition of unemployment and the labor supply (Table 3.13). The demand for unskilled labor drops more rapidly

Table 3.13 Employment and labor supply by level of education.

Level	Employment (perc.)			Supply (perc.)		
	1975	1985	2000	1975	1985	2000
elementary	23.8	14.1	7.7	24.9	16.2	10.6
extended elementary	30.8	28.2	23.6	31.0	29.3	24.3
secondary	32.9	39.9	45.5	32.2	37.9	42.5
semi-higher	9.0	13.2	17.3	8.7	12.3	15.8
higher	3.4	4.7	5.9	3.2	4.3	6.9
Total	100	100	100	100	100	100

Source: Kuhry & Van Opstal 1988, 73-5

than the supply declines, with the consequence that unemployment strikes primarily this category. The demand for mid-level and higher employees shows a relatively strong growth.

Up to this point, the argument supports the idea of a relatively reduced role for industry, but it has relied solely on the criterion of employment. A more complete picture of the place of industry in the economy can be sketched with other indicators. A comparison of the sectorial contribution to employment and to the gross value added of firms is revealing: a relative gain of employment in the tertiary sector does not necessarily imply that the gross value added also increases (Figs. 3.8, 3.9). The degree of labor or capital intensity of the various sectors offers an explanation for this contradictory development. In the crisis that started in 1974, industry watched its relative contribution to the gross value added fall much more abruptly than its proportion of employment. In this context, it is noteworthy that in the period 1982-1986 the gross value added (at market prices and with the spending-power of the guilder in 1982) in industry rose continually in absolute terms. The value in 1986 was 14.2 percent higher than in 1982. In the same period the volume of labor decreased by 4.0 percent.

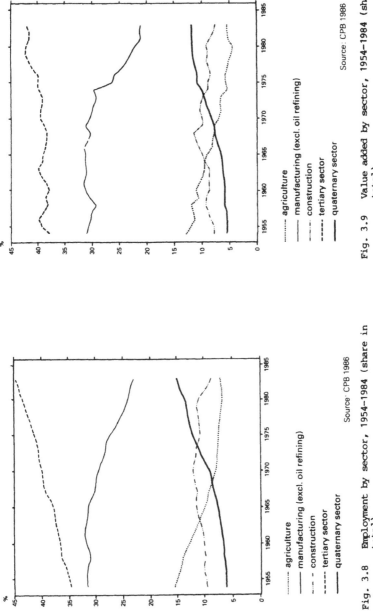

%

agriculture
manufacturing (excl. oil refining)
construction
tertiary sector
quaternary sector

Source: CPB 1986

Fig. 3.8   Employment by sector, 1954–1984 (share in total)

%

agriculture
manufacturing (excl. oil refining)
construction
tertiary sector
quaternary sector

Source: CPB 1986

Fig. 3.9   Value added by sector, 1954–1984 (share in total)

A long-term comparison of the mean levels of economic indices in industry relative to other sectors shows that some trends apply to the entire economy, and others, in contrast, are only characteristic of the industrial sectors (Table 3.14). Thus, the income quote for labor rises across the board and the investment quote falls (in percent of the gross value added). Sectors other than industry show a steeper investment curve. Government investment was much greater in the 1960s than in the 1970s. No wonder government investment (particularly in infrastructure) was recently cut back to the 1963 level. The enormous investments in the sixties in oil refineries, for instance, dropped sharply after the oil crisis. Agriculture, on the other hand, has experienced an increased investment input (intensification). Industry's share in production (value added) and employment confirms that it is in relative decline (from approximately 30 percent in the 1960s to less than 25 percent during the oil crisis).

A number of other indicators underscore the difference in dynamics between industry and the other sectors during periods of boom and crisis (Table 3.15). Industry has continually shown a significantly greater rise in productivity than construction and services, in part because the latter have more access to factor substitution (replacement of labor by capital via mechanization and automation). The rise in productivity in agriculture was maintained after 1973 at a high level; this was not the case in industry. In particular, production volume (value added) showed only a

Table 3.14 Key figures by industrial sector, 1963/1973 and 1973/1983 (average levels).

	agri-culture	manufac-turing[3]	energy	construc-tion	ter-tiary	quater-nary	total
1963/1973							
labor income quote	74.8	73.8	49.3	77.1	81.8	75.0	78.0
investment quote[1]	16.6	18.0	66.7	7.4	15.8	12.9	18.5
production share[2]	7.6	30.4	5.2	10.0	38.8	7.8	100.0
employment share	8.7	29.8	2.0	11.5	39.2	8.8	100.0
1973/1983							
labor income quote	86.3	88.0	17.9	87.8	89.9	81.2	88.5
investment quote[1]	26.4	13.9	20.2	6.1	13.1	5.8	14.7
production share[2]	5.2	23.7	10.1	8.5	41.1	11.4	100.0
employment share	7.1	25.1	1.6	10.4	42.6	13.2	100.0

[1] Percentages of gross value added, at current prices, investment by destination, housing excluded.
[2] Value added (factor costs).
[3] Oil refineries excluded.

Source: CPB (CEP 1986). For 1953/63 see Table 3.10

Table 3.15   Key figures by industrial sector 1963/73 and 1973/83 (annual percentage fluctuation).

	agri-culture	manufac-turing[2]	energy	construc-tion	ter-tiary	quater-nary	total
1963/1973							
labor productivity	7.8	7.1	16.8	2.4	3.8	-1.0	5.4
volume of production[1]	4.5	6.4	12.3	3.5	5.1	3.9	5.9
volume of consumption	3.4	4.8	7.4	0.7	5.4	5.6	5.4
volume of export	6.2	11.6	14.2	--	6.0	--	10.3
1973/1983							
labor productivity	6.2	3.9	-0.1	0.4	2.4	0.6	2.2
volume of production[1]	4.9	1.1	0.1	-3.0	2.6	3.1	1.5
volume of consumption	2.6	1.2	1.7	-3.6	3.2	3.4	2.1
volume of export	6.2	3.1	-2.0	--	2.4	--	2.7

[1] Value added (factor costs).
[2] Oil refineries excluded.

Source: CPB (CEP 1986)

minimal increase in industry after 1973, in contrast to agriculture. Data on consumption and export leave no doubt about the real cause. Industry has had to deal with a much more resistant stagnation of its international market than agriculture, with its strong export orientation. Fig. 3.10 illustrates how deep the gulf was that industry had to traverse. The level of investment around 1982 was not even half that of 1970. Meanwhile, recovery has set in.

Industry has thus made a poor showing across the board since 1973 in comparison with other sectors. This of course does not imply that industry has declined in absolute terms. The question is whether in this respect great differences will appear within the industrial sectors.

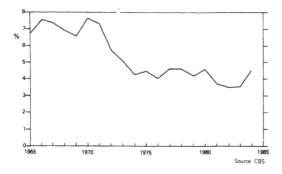

Source CBS

Fig. 3.10   Investment in the manufacturing sector as percentage of annual sales, 1965-1984

Dynamics Within the Industrial Sectors

Three trends may be distilled from the overview of the industrialization of the Netherlands in the period 1953-1967: a strengthening of basic industry, a rise in the share of industry groups employing skilled labor and an increase in the scale of production (Jansen *et al.* 1979: 57 ff). After 1967, a thorough restructuring process took place in many industry groups and the increase in scale was due to mergers and take-overs. The basic industry - including steel mills, oil refineries and the petrochemical industry-the highly specialized metal manufacturing firms (including machinery and electro-technical products) and the printing industry were the growers. In 1953 these accounted for only one-third of all industrial employment in the Netherlands; by 1971 their share had grown to half.

Yet there were also weak industry groups, such as the cotton, rayon and linen industry and footwear manufacturing, both of which shed half of their jobs in the same period. This was a hard blow to the areas where these sectors were concentrated (Twente, Tilburg, Helmond, Langstraat). The situation was even worse in shipbuilding (Amsterdam, Rijn-mond) and coalmining (Zuid-Limburg). It is remarkable that the apparel industry, which had continued to grow for a long time in the Netherlands due to the low wages in this sector, did not depart for foreign locations until around 1970 in order to survive.

The causes for the decay of these industry groups should be sought primarily in the low elasticity of demand for the articles produced, the rise of substitute articles and increasing competition from the cheap-labor countries. In addition, the middle-sized, family-owned companies were subject to weak management (Van Gelder 1973).

Comparison of the trends of the 1950s and 1960s with those that emerged later is complicated by the inconsistency of the statistics on industry groups. It is clear, however, as pointed out above, that before 1972 the Netherlands made a good showing internationally in terms of production growth, but afterwards did rather poorly. The Netherlands demonstrated a stronger growth in internationally expanding industry groups (e. g. chemicals) at the peak of its economic growth - due to its delta location (ports) - and rational-ized the weak industry groups more rapidly during the oil crisis (hastened by the small domestic market and the greater dependency on the international market).

Even before the economic recession, which started around 1972 and was deepened by the oil crisis at the end of 1973, Dutch industry was already embarking on a process of restructuring. Although this had painful repercussions on the regions, there was initially enough compensation at the

national level in terms of employment. This changed, however, after 197ₗ. The availability of comparable data only
allows comparison of the dynamics for the major categories
of industry groups over the last 10 years in terms of

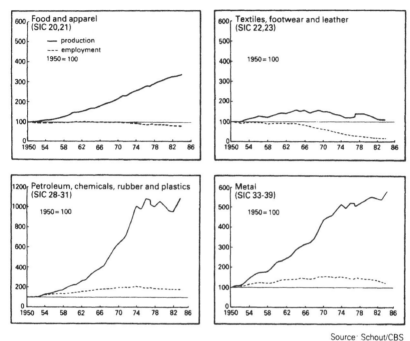

Source· Schout/CBS

Fig. 3.11 Production and employment in selected sectors,
1950–1984

employment, sales volume, growth and export orientation
(Table 3.16). Fig. 3.11 gives some insight into the
production volume by comparing the development of some
characteristic industry groups (1950-1985) with the
development of employment.

Employment in the textile industry and clothing firms
has undergone a drastic decline during the past 10 years
(to 50 and 41 percent of the jobs in 1976!). Other sectors
that proved to be weak were leather goods and footwear
manufacturing, which was to be expected, and the lumber and
furniture industry, as it was closely related to the crisis
in the construction industry. Growth of employment was
limited to the rubber and synthetic fiber industry, electro-technical manufacturing (recent recovery of Philips)
and the service-oriented publishing houses.

Further analysis of the statistics on production volume
reveals that although the growth rapidly tapered off after

Table 3.16 Number of jobs, increase of sales and share of export by industry, 1976-1985.

Industry	Jobs (x 1000) 1976	1985	1985 (1976=100)	Sales increase 1976-1985 (industry=100)	Export as perc. of sales 1976	1985
Food and kindred products	148,4	134,1	90	99	28	37
Textiles	46,8	23,2	50	69	44	51
Apparel	26,0	10,6	41	41	24	34
Footwear and leather goods	9,3	6,0	65	76	21	34
Lumber, furniture	36,9	24,0	65	67	12	13
Paper	29,1	23,2	80	105	35	45
Printing, publishing	59,5	61,3	103	101	10	12
Oil refineries	10,1	9,7	96	109	53	47
Chemicals, fibers	89,3	84,4	95	127	67	74
Rubber, plastics	24,8	25,6	103	114	35	44
Building materials	37,5	29,2	78	90	19	23
Metallurgy	39,2	30,6	78	92	64	71
Metal products	89,9	70,4	78	90	25	29
Machinery	88,0	78,8	90	99	49	50
Electronics	113,1	115,7	102	189	65	61
Transport equipment	80,8	60,5	75	124	57	58
car, aircraft	21,0	21,4	102	153	72	82
shipbuilding	47,4	25,6	54	90	55	48
Precision and optical instruments	9,6	7,2	75	75	69	46
Other	5,2	4,3	83	137	22	23
Total industry	943,6	807,7	86	100	43	48

Source: CBS, Alg. Ind. Statistiek

the 1960s few major industrial groups have shown an absolute decline in production since then (Fig. 3. 11). The exceptions are the old standbys like textiles, leather goods and footwear and, after 1970, apparel as well. The decline was actually much less than could have been expected on the basis of the decrease in jobs. Factor substitution also took place in these industry groups, and very strongly in the cotton industry. Recovery of production made headway in the 1980s, particularly in textile and footwear manufacturing. The petroleum industry, construction materials firms and shipbuilding have a lower level of production. In some branches, the recent growth in production is even startling, as in the paper industry (after restructuring with a heavy loss of jobs), the electro-technical industry, and in tool-making (though with less growth of employment than anticipated).

The results of analysis depend on which indicator is chosen. Analysis of statistics on the growth in sales, for example, rounds out the picture. It should be kept in mind that prices develop differently per industry group, as, for instance, demonstrated by the influence of oil prices on the chemical and related industries. The background of the differential developments has been sketched for the weak parts. The basic industry (metallurgical and petroleum industries) has felt the backlash of the economic crisis. There is overcapacity in these sectors worldwide. The industry groups that enjoyed a more advantageous sales position were usually involved in export (Table 3. 16). This is well illustrated by the paper industry, food and kindred products, and the chemical industry. In the weak sectors, the higher share of export in apparel and footwear, for example, is an indication that the sales stagnation was mainly domestic. The high rate of exchange for the US dollar at the time exaggerates the share of export in the total sales.

Large and Small Industries in the Netherlands

The structure of Dutch industry can be described not only with economic indicators but also by the size of operational units. The sixties were characterized by functional concentration. Large firms grew more rapidly than small ones and even swallowed them up through mergers and take-overs. Yet in the last 10 years there have been signs that small scale is going to gain importance in industry. The term 'small and medium-sized enterprises' (SME) should be used with caution. The official definition refers to private businesses (not firms!) with fewer than 100 employees. In other countries the upper limit is mostly higher, and the term is often applied to enterprises with a maximum of 500 employees (Nooteboom 1986). Adoption of this high ceiling would ex-

Table 3.17 Manufacturing plant size at year end 1974 and 1983; number of plants and jobs.

Size category (jobs)	Number of plants			Jobs (x 1000)		
	.983(end)	1974(end)	1974-83 perc.	1983(end)	1974(end)	1974-83 perc.
10 - 19	3,365	3,285	+ 2.4	49,2	48,1	+ 2.3
20 - 49	3,072	3,357	- 8.5	95,4	107,0	-10.8
50 - 99	1,274	1,624	-21.6	86,7	114,1	-24.0
100 - 199	720	920	-21.8	95,1	128,5	-26.0
200 - 499	422	576	-26.8	121,0	173,8	-30.4
500+	219	277	-20.9	336,9	435,1	-22.6
Total	9,072	10,039	- 9.4	784,4	1.006,6	-22.1

Source: CBS, Maandstatistiek van de industrie

clude only 2 percent of the Dutch businesses at most! Limiting the number to a maximum of 100 employees implies that the SME sector would account for a quarter of the national production, a share to be reckoned with.

The category of small and medium-sized enterprises should be differentiated. The breaking point is ten employees. The small business predominates in the trade sector, where only one out of eight establishments have ten or more employees (many small independent businesses). In construction, three-quarters are small businesses, and in manufacturing and crafts, almost two-thirds (17,000 in 1985). However, the small manufacturing businesses, often craft-like in character, are not included in the employment summaries of the general industrial statistics, which are repeatedly used as sources of data in this book (Algemene Industrie Statistiek). Table 3.17 shows that the medium-sized enterprises (10-99 employees) accounted for 85 percent of the total number of manufacturing companies at the end of 1983, but less than 30 percent of the employment. Half of the medium-sized enterprises employ between 10 and 19 persons. This category provides only 6 percent of the total employment in industry. The large industrial company, with 2 percent of the number of businesses, provides 43 percent of the industrial employment. That share has been maintained throughout the economic crisis (since 1974).

The weak link in Dutch manufacturing appears to be the family-owned businesses with between 200 and 500 employees (Table 3.17). These businesses in particular were concentrated in industry groups that were hard hit. Large and small enterprises are unevenly distributed across the industry groups (Table 3.18). In the metallurgical and petrochemical basic industry, the large business predominates, mostly as part of internationally operating corporations, with or without a Dutch background. The same applies to electro-technical manufacturing (Philips) and the transport vehicle

Table 3.18 Employment by some industrial sectors and size of plant, at year-end 1974 and 1983 (x 1000).

| industry | End 1974 | | | | End 1983 | | | |
	small (10-49)	medium (50-199)	large (200+)	total	small (10-49)	medium (50-199)	large (200+)	total
food and kindred								
products	21,7	35,1	96,8	153,6	24,4	30,8	80,3	135,5
textiles	4,8	14,8	35,5	55,2	4,5	8,1	10,9	23,4
apparel	8,6	11,1	13,4	33,1	4,1	5,7	2,0	11,8
footwear and								
leather goods	3,8	7,2	0	11,0	2,5	3,0	0,9	6,3
lumber, furniture	17,2	15,8	6,9	39,8	13,7	8,5	2,5	24,7
paper	2,6	10,0	19,0	31,6	2,1	8,4	12,5	22,9
printing	16,0	20,9	24,6	61,4	16,7	17,5	25,4	59,4
chemicals	4,4	15,9	73,5	93,8	4,2	9,9	67,2	81,3
metal products	25,1	29,7	41,1	95,8	22,8	24,2	20,6	67,5
machinery	18,3	28,7	45,9	92,9	19,0	25,1	30,3	75,0
transport equipment	9,3	14,6	59,3	83,2	7,5	9,2	45,0	61,7
industry (total)	155,1	242,6	608,9	1006,6	144,6	181,8	457,9	784,4

Source: CBS Maandstatistiek Industrie

industry (Fokk__, DAF, Volvo, large-scale shipbuilding). Businesses in sectors such as apparel and footwear are small. As in furniture manufacturing, these small businesses have had a hard time in the crisis period and had to pay dearly in employment. In the textile industry a number of large companies fell by the wayside (the number of companies with 200 or more employees dropped in the period 1974-1983 from 72 to 28!). Sometimes small units are left, as a result of management buy-outs, for instance.

There is a general trend toward functional deconcentration. The small businesses and the category of relatively small, medium-sized enterprises (10 to 19 employees) are reasonably able to hold their own (Table 3. 17). Should small scale be considered positive in all senses? We will approach this question from various perspectives, namely in terms of economic indicators, position in the product life cycle, export orientation, and functional connectivity with other sectors of industry.

As the business expands, industry in general sees its productivity rise (this also applies to the value added). Comparison of two industry groups, on the other hand, shows that this principle does not always apply (Verhoeven & Vianen 1984). It is valid in the case of metal manufacturing, but not for food and kindred products (Fig. 3. 12). In the latter sector, economies of scale proved to be greatest in medium-sized enterprises with 50 to 99 employees. Small businesses, usually craft-oriented, in both industry groups lack economies of scale. Research to elucidate the background of these economies of scale has yet to be carried out.

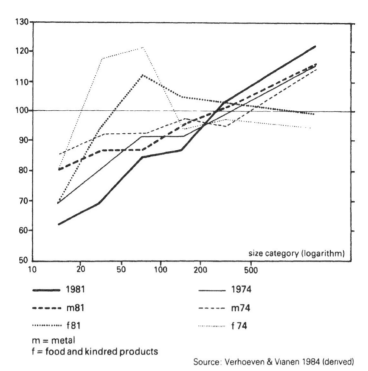

1981 ——— 1974 ———

m81 ----- m74 -----

f81 ·············· f74 ··············

m = metal
f = food and kindred products

Source: Verhoeven & Vianen 1984 (derived)

Fig. 3.12 Production per employee by firm size for all
manufacturing, in the food and kindred products
sector and in the metal sector, 1974 and 1981

Table 3.19 Large and medium-sized manufacturing firms according to
four indicators for each life cycle stage.

life cycle stage	indicator*			
	industrial renewal	market orientation	rate of return	activity
expanding	l	l	l	l
slightly stagnating	l	l	m	l
stagnating	m	l	m	l
saturated market	m	m	m	m
slightly receding	m	m	m	m
strongly receding	m	m	l	m

* m = medium-sized firm scores higher than large firm
  l = large firm scores higher than medium-sized firm

Source: Webbink 1985b

A different perspective on the position of the medium-sized business (10 to 99 employees) is the place in the product life cycle. The context of the medium-sized business may be outlined in the following points (Table 3.19, Webbink 1985b):

1. The big business in industry groups with a strong expansion is in a better position than the medium-sized business.

2. The medium-sized enterprise makes a better showing than the big business within industry groups that are in a relatively poor position (the smaller business in the role of buffer).

3. The big business thrives in situations with impediments to growth; smaller enterprises are more adept at utilizing special submarkets (niches) and even manage to realize innovations in the less expansive parts of the industry.

It should be added that research and innovation are less feasible for the SME sector than for the large corporations. The financing structure is also weaker, unless a company is incorporated in a larger business context. No less of a problem for the SME sector is the lack of sales markets, particularly in foreign countries. It is nonetheless encouraging to observe that, in terms of export growth, recently small and big business in industry have been showing the same dynamic. However, major differences appear between industry groups (Table 3.20). The impending restructuring in the large paper manufacturing firms has had an impact on the stronger export position; in the electro-

Table 3.20 Growth of industrial export by establishment size and industry group, 1976-1983 (in perc.)*

industry group	medium-sized establishment (10-99 employees)	large establishment (100 or more employees)
food and kindred products	65	96
textiles, apparel, leather goods	110	-29
lumber and construction materials	51	55
paper	12	82
printing	93	51
chemical and synthetic products	110	101
metal	56	33
all industry	68	67

* In current prices

Source: EIM

technical industry and parts of the food and kindred products sector, the big corporations have contributed to a forceful growth in export. Yet it is striking how the strong export growth of the SME in weak sectors like textiles, apparel and leather goods contrasts with the retreat of big business on the international market. The crisis in shipbuilding and the (related) machinery industry manifests itself in big business as a low level of growth in export (measured in non-indexed prices!). Small, specialized enterprises have nonetheless been able to find new markets. This apparently demonstrates that 'small is beautiful', and big business proves to be too cumbersome of an entity.

Finally, the question may be posed how the SME sector can fulfil an optimal function within industry. In Japan the SME sector is part of an outer ring in the economy. It offers relatively low wages and weak job security to the personnel. Moreover, it is entirely dependent on and, thus, highly sensitive to conjunctural fluctuations, since big business subcontracts work to enterprises in this sector. It

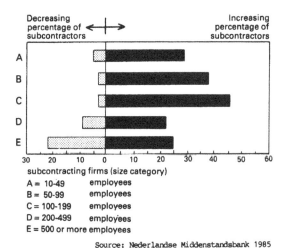

Source: Nederlandse Middenstandsbank 1985

Fig. 3.13 Changes in the process of subcontracting,
differentiated by firm size

is not advisable for the Netherlands to follow this example. Instead, it is better to learn from the Japanese situation in regard to flexibility (by refusing to grow) and complementarity (co-makership). Despite the fact that the big corporations are cutting back drastically on the number of subcontractors (e. g. Océ van der Grinten, from 5000 to 350), there is still no one-sided dependency on suppliers, as in Japan (Fig. 3. 13). Only one out of every eight entrepreneurs in the SME sector is dependent for at least 70 percent on three big contractors. The SME sector should give more attention to building up good networks, in order to fulfil a flexible, complementary function among the big corporations. Reciprocal exchange within the SME is clearly increasing in importance, which allows creation of a division of labor with economies of scale (NMB 1985).

3.4 Industrial Structure Policy

In the previous section, changes in the industrial structure were analysed on the basis of various criteria. As Van Rhijn (1971) determined, industrial structure policy is concerned with all sectors. It is not limited to the rescue of eroding industry groups. The question arises whether there is a sequence of generations of sectorial structures (Van der Zwan 1980). In Japan (MITI) and South Korea, goals have been formulated in regard to priorities for specific industry groups in subsequent periods. The developmental sequence runs from labor-intensive (with relatively low wages) via capital and energy-intensive (basic industry) to high-skill technological activities. Naturally this cannot be the only perspective. Profitability and growth potential should also be kept in mind, certainly in regard to an economy like that of the Netherlands.

There is some debate on whether an industrial structure policy is feasible in the Netherlands. A fervent opponent like H.W. de Jong characterizes the Dutch structure policy as 'the visible finger on the invisible hand' and as 'an empty box' (De Jong 1980, 1985a). In his view, the Netherlands is too small and too open, a delta economy at the crossroads of large industrial nations, which greatly limits the room for an autonomous policy. 'Structure policy of the Dutch government in those branches of industry where large multinational corporations already exist, or where cartels operate, or strong cooperatives are found, must certainly imply sanctioning that which the entrepreneurs have already decided', according to De Jong (1980: 49-50). He concludes that 'here is the manifestation of the other, sometimes hardly perceptible, finger that goes with the hand of the market process'.

The Memorandum on selective growth (1976), subtitled Economic structure memorandum, argues for a sector policy within the framework of an oriented market economy. 'Oriented' refers to the (limited) possibility to steer the direction of economic development in an open market economy. The two main objectives in this government memorandum, continuity and selective growth, imply that the sector structure (composition of industry groups) should make as smooth an adjustment as possible, in order to conform better to the role of the Netherlands in the world economy. This process of adjustment should take account of goals regarding use of resources, environmental protection, physical planning and international division of labor. The bill (for environmental degradation and damage to physical planning) for an unfette- red economic growth had not yet been paid at the time. The energy crisis and the collapse of industry groups sensitive to wage costs (e. g. textiles, apparel, footwear) due to international competition compelled the policy-makers to reconsider the situation.

It is definitely no easy task to formulate a structure policy. In fact, the 1976 memorandum outlines what structure policy could embody, taking into account t.ie narrow action radius of an open economy in a small country incorporated in the large European Economic Community. A structure policy can be defensive and/or offensive, but it always sets the conditions for the improvement of the production environment (Wijers 1985, 182). The restructuring of traditional indus- try groups has already been discussed. This policy was intensified and had a defensive character. Priority was given to retaining jobs. An offensive policy for manufac- turing activity with good prospects and of a high technolo- gical level was indeed announced (so-called spearheads), but it was not detailed. The subsequent sector memorandum (1979) fulfilled the role of a 'progress memorandum in reference to economic structure policy', and its major aim was to stem the tide of financial support for individual businesses which were on the brink of failure. An impulse for a new way of thinking about structure policy was not felt until the report on the position and future of Dutch industry was published (WRR 1980). The importance of this report for a view of the international position of the Netherlands was sketched above. The blueprint for the restructuring should give some additional attention to the ideas presented in this report.

The WRR report observes that the pattern of (indus- trial) specialization of the Netherlands is under competi- tive pressure from four sides: for advanced products, from the side of the US, Japan and West Germany; for capital goods, from the side of the Newly Industrializing Countries (NICs); for petrochemical basic industry, from the side of the OPEC countries; and for labor-intensive products, from the side of developing countries. Import statistics show that the penetration of the Dutch market by foreign sup- pliers grew in the 1970s in virtually all industry groups, including electronics and transport vehicles. All in all, there are several structural problems that require a sol- ution. These include the rise in wage costs, the one-sided composition of the Dutch export package (high in bulk chemi- cals, low in capital goods), the revaluation of the guilder (a hindrance to export), and the one-sided geographic orien-

tation of the export (more than 80 percent to Western
Europe, whereas the economic growth was taking place elsewhere).

What are the options? The WRR report (1980) advised
'making visible the contours of the future economic structure, to which private parties and government can effec-
tively react' (p. 16). The preference is not for an exclusively defensive policy (as for instance in shipbuilding,
where unprofitable projects were financed with a great
amount of government funds). But this does not imply that
the plight of weak industry groups should be disregarded.
Revitalization should be given top priority. There should be
a concerted effort to find higher-quality products and new
markets for these industry groups as well. The emphasis on
bulk petrochemical products should be dropped in favor of
more specialized products that are higher in the industry
column, hence closer to the consumer. Besides these two
elements, a dynamic import substitution is suggested as a
third tenet of an industrial strategy. This is aimed at
strenthening the equipment sector (machines, engines, heavy
electro-technical equipment), which is weakly represented
in the Netherlands, due to the obvious disadvantage of a
limited domestic market.

The WRR report advises turning the advancing process of
de-industrialization into re-industrialization, a path that
could lead to a new and appropriate pattern of industrial
activities. In the four reports by the advisory committee on
industrial policy (the above-mentioned 'Wagner Commission'
1981-1983), this path is further specified. Briefly, its
core is a two-track policy: the creation of good conditions
(control of wage and energy costs, training) and structure
policy (rationalization of the subsidies for individual
enterprises, re-industrialization). To achieve a stronger
structure, it suggests making a selection of high-potential
activities as well as formulating a government acquisition
policy and actually taking over establishments of foreign
firms. Big businesses that are in financial straits can
consult an advisory council for the financing of economic
recovery. This agency, in operation since 1983, is comprised
of independent experts who evaluate the options for support.

In practice, industrial policy in the Netherlands has
not had a good press. The government had to foot the bill
for the immense losses of enterprises in weak industry
groups such as textiles and shipbuilding (the RSV scandal
being the most excessive). The investment subsidy act (WIR),
intended to promote investment, was too widely applicable to
buttress the structure policy. Neither micro nor macro
policy provided a solution, but policy at the intermediate
level met with equally sharp criticism. In 1973 the Netherlands Restructuring Corporation (NEHEM) was established to

implement the sector policy through tripartite consultation (government, employers' organizations and trade unions). A common strategy was expected to emerge from the negotiations of these parties for directed growth or contraction, with special attention for concerted efforts in export, harmonization of capacity, and technology (research and training). Government and business have insufficient access to information to do this. Intermediate policy would thus form a link between macro and micro policy (Wijkstra 1979).

The crux of implementing a sectorial structure policy at the intermediate level is that 'there are no strong and no weak branches, just strong and weak businesses' (De Wolff 1983: 23). Within the NEHEM, there is a strong tendency to steer toward horizontal relations. Too little interest is shown in diagonal relations of cooperation between businesses that are related but not identical. In a defensive mode, there is a drive toward mergers, whereas a smaller scale in business is conducive to a smoother adjustment (Schenk 1985). A study of shipbuilding revealed that a business has a better chance of survival when it is embedded in a network rather than in a large, isolated firm. When small businesses are involved, the division of labor is clear and the cost controls more lucid (De Feyter 1982).

Despite the options for a more creative intermediate-level policy, the problem remains that structure policy - in addition to the above-mentioned points over the open and small domestic economy - is only capable of restructuring the supply side, not the demand exercised by the market. In large and less 'open' industrial nations there may be a 'bureaucratic symbiosis', a concerted effort of the elites in government and the business community, aimed at government acquisition of technologically high-quality but expensive products, primarily in the first stage of the product life cycle (e. g. in France: Groenewegen 1985).

As mentioned above, the target area policy (the so-called Wagner list, Table 3. 7) attempts to pinpoint the spearheads of industry in the Netherlands. It is important to be aware that 'a weak point is the defective interconnection of branches of industry' (e. g. the inadequate connection of the metal industry to processing activities). Moreover, it is 'an illusion to think that the Netherlands could become a true first-to-market nation in the near future. It is better to concentrate on innovations in the international trade, transport and service sector' (Franke & Whitlau 1979: 189). This opinion is shared in the main by De Jong (1980).

In the long run, the question remains whether the growth of the high technology equipment sector would have borne fruit after all. In 1958, this sector had an 8 percent share in business investments in fixed assets, 14 percent in 1973, and an impressive 33 percent in 1983 (specifically in

office automation). These figures are, however, somewhat
exaggerated due to the decline in investment in business
plant (Van Duijn 1985: 48). In more general terms, the
course set out in the reports by the Wagner Commission and
elsewhere may be characterized as follows: 'the exploration
of thin markets for specialized products, whereby the production characteristics protect the market more than the
price does' (Van Duijn 1980: 76). Even though the industrial
structure policy has been firmly criticized, and rightly so,
it is a misjudgment of Dutch industry to sketch the future
of the Netherlands as the Manhattan of Europe (Franke &
Whitlau 1979: 185) or to compare the country with the city-
state of Singapore. Policy has entered an offensive mode.
The emphasis lies on policy directed toward leading enterprises with combinations of product, market and technology
that have good prospects, often unique in their kind in the
Netherlands (Philips, Fokker, DAF, etc.) and thus operating
without unfair domestic competition (Wijers 1985: 188). The
question is, which effects will remain within the Netherlands? And to what extent can the small and medium-sized
enterprises profit from these? Another way to finance large
projects with good prospects for the future is through the
Corporation for Industrial Projects (MIP), which operates on
a modest scale, for the time being, as a partnership com-

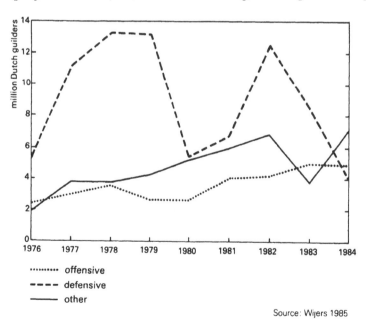

Source: Wijers 1985

Fig. 3.14 Government subsidies in manufacturing;
offensive and defensive strategy, 1976-1984

pany. It may become an important source for funding developments that have been foreseen in the target area policy.

A change in industrial policy has been underway for the past 10 years. The defensive policy, i. e. the tendency to keep traditional business activities alive, mostly by artificial means, is being phased out and the offensive policy is gaining momentum (Fig. 3. 14).

4 Regional Industrial Patterns and Trends

4.1 Trends in Location and Regional Specialization: The Industrial Pattern Around 1930

There is a complex relationship between the spatial pattern of manufacturing and structural changes in the economy. The region offers options and poses locational constraints on industry. Since the configuration and the attraction of the locational factors are not regionally uniform, manufacturing differs remarkably per region. This configuration of factors and the amenities form the undercurrent of the structural changes in a region's economy.

The economic-geographic picture of the Netherlands was long dominated by regional specialization, particularly in the period 1930-1963. This industrial pattern deserves some consideration before going more deeply into its development. How did it evolve and how is it connected to long-term structural economic changes at the national level?

Even though the pattern of regional specialization did not appear everywhere at the same time, location theory can elucidate a number of trends in industrial activity. The situation around 1930 still lends itself to explanation in terms of the main principles put forward in the classical theory of Alfred Weber (1909). This perspective reduces locational choice to a weighing of transport costs (cost-minimizing behavior). In that era of predominantly water and rail traffic, transport was the prime factor in the location decision.

Weber's theory conceded that regions with an unfavorable transport location could attract firms if the extra cost of transport were compensated by a lower wage rate. The points where these costs are in balance fall within the bounds of what Weber called the critical isodapane. Compensation could be provided via agglomeration economies (Krusel 1957) as well as in labor costs. Such economies are found when similar production units converge in one area on the basis of mutual relations (locational economies) and/or when essentially dissimilar establishments, such as educational and communication facilities, business services, etc., have

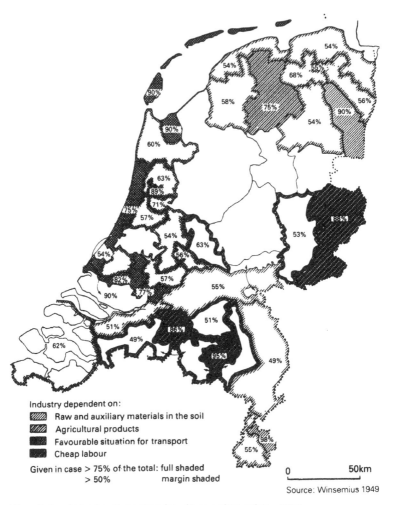

Fig. 4.1   Locational tendencies in manufacturing, 1930

common locational requirements (urbanization economies).

On the basis of these theoretical notions, Winsemius (1949) constructed a diagram of the locational factors operating in the manufacturing sector in 1930 (Table 4.1, Fig. 4.1). First, he differentiated basic and non-basic factors. The latter type of factor, on which Winsemius did not elaborate, applies to businesses that have no supra-regional market to speak of. An essential difference is found between primary and supplementary basic factors. The primary basic factors (the so-called core factors) are derived from Weber's emphasis on transport (in the case of dependency on raw materials and transport situation) or from

Table 4.1 Typology of locational tendencies in manufacturing, by
employment (1930).

Locational tendency	Employment
	1930 (31-12)
A. basic tendencies	640,800
A-a original basic tendencies (nucleus tendencies)	
A-a-1 dependent on raw and auxiliary material in the soil	87,900
A-a-2 dependent on agricultural products	49,400
A-a-3 dependent on favorable situation for transport	139,700
A-a-4 dependent on cheap (male adult) labor	152,200
	429,200
A-b derived basic tendencies	
A-b-1 dependent on cheap female and juvenile labor	48,900
A-b-2 dependent on skilled labor	38,900
A-b-3 auxiliary industry	37,700
A-b-4 dependent on the city	59,400
A-b-5 dependent on the market	26,600
	211,600
B service tendencies	595,100
unclassified	100
total for manufacturing and mining	1,235,900

Source: J. Winsemius 1949, 18

his focus on labor. The agglomeration principle pertains to
the supplementary basic factors (presence of auxiliary
manufacturing). The availability of unskilled female and
juvenile labor depends on whether male breadwinners are
present. It is noteworthy that Winsemius classifies the
availability of skilled labor among the supplementary basic
factors (an urbanization economy). The economic base of a
region is determined primarily by the core factors (Fig.
4.1).

   An economic-geographic picture of the Netherlands was
sketched in terms of industrial (sub)sectors. These were
classified on the basis of their reliance on either primary
or supplementary basic factors (while acknowledging variations in the strength of dependency). Limiting the discussion to the core factors depicted in Fig. 4.1, the extent to
which a given locational factor dominates the interregional
locational preference for a given region can be determined.
In the West the transport situation dominates among the
location factors. In the peripheral areas of the East and in
the southern province of Noord-Brabant, the composition of
the labor force was the major factor. In the North, the
availability of agricultural resources predominated. And in

the central riverine districts and the mining region of Limburg, mineral extraction prevailed. The regions differ, however, in terms of their dependency on these core factors. Furthermore, it should be realized that in the large metropolitan areas, these supplementary basic factors (urbanization economies) are essential to the creation of employment. Basic (private sector) services, which were not treated by Winsemius, were also strongly oriented toward the large metropolitan areas; many cities outside the metropolitan areas find a major source of revenue in the provision of regional services. Hence, the mosaic of regional specializations is comprehensible. This pattern book remained essentially the same until 1963, though with a few alterations.

4.2 Industrial Restructuring and Regional Employment

The transition from industrial growth to restructuring, described in Chapter 3, also applies to the industrial development of the regions. Reconstruction generally has specific regional repercussions. Around 1963, it was easy to identify industrial regions on the map of the Netherlands that had retained their own particular characteristics. In some instances, only one type of manufacturing, such as textiles, determined the character of an area. It is in regionally concentrated manufacturing sectors that restructuring took its greatest toll. The traditional industrial-geographic pattern has lost much of its meaning. The mosaic of industrial areas, each with its own character, has become blurred. Manufacturing is losing ground virtually everywhere in the country as the motor of employment, though the degree varies strongly by region.

The industrial dynamics are first presented in terms of employment figures. Subsequently, other indicators are considered (4.3). Rationalization and automation may have given a new lease on life to firms where job losses had been heavy. Some trends in long-term regional industrialization, from 1963 to the mid-1980s, will be treated first. Then the absolute and relative importance of the manufacturing sector for the regions will be discussed. Finally, the relation between manufacturing sectors and regions will be elaborated; in other words, to what extent does regional concentration still take place and where do weak or strong aspects of production emerge?

Regional Contrasts in Industrial Change, 1963-1985

The industrialization of all regions was a clear success in the 1960s. The most spectacular manifestation of its achie-

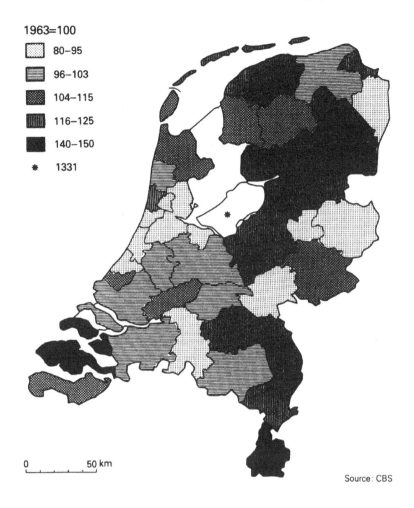

1963=100

▨	80–95
▤	96–103
▨	104–115
▥	116–125
■	140–150
✳	1331

0 _____ 50 km

Source: CBS

Fig. 4.2   Regional development in manufacturing,
          1963–1973 (employment)

vements was in the ports outside the West (Fig. 4.2), which
profited from the limited capacity in Rotterdam. The harbors
along the River Scheldt, which have the added advantage of
being located on the water approach to Antwerp, especially
benefited from the growth of metallurgical and petrochemical
industries. But there was growth elsewhere as well. Hoogovens expanded its steel plant at IJmond, located at the
entrance to the North Sea Canal. A number of lagging areas,
including parts of the provinces of Friesland and Drenthe as
well as regions in the southeastern part of the country,
were also able to attract new manufacturing firms. Here,

71

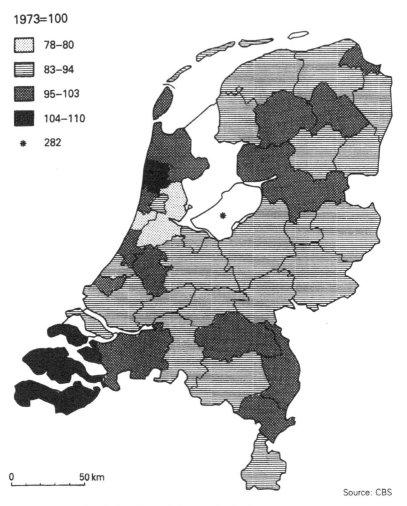

1973=100

▨	78–80
▤	83–94
▨	95–103
▨	104–110
✳	282

0   50 km

Source: CBS

Fig. 4.3   Regional development in manufacturing,
           1973-1979 (employment)

subsidiaries of large domestic and foreign corporations
found the labor that had become unavailable in the western
part of the Netherlands. Labor market considerations promp-
ted some establishments to move. In most cases, the reloca-
tions benefited the areas adjacent to De Randstad: the
northern part of the province of Noord-Holland, the Veluwe
area, and the western part of the province of Noord-Brabant.

There have been major cutbacks in industrial employment
since the mid-1960s. First hit were textiles in Twente,
coalmining in Zuid-Limburg, and textiles and footwear in
the central part of Noord-Brabant. Restructuring also made

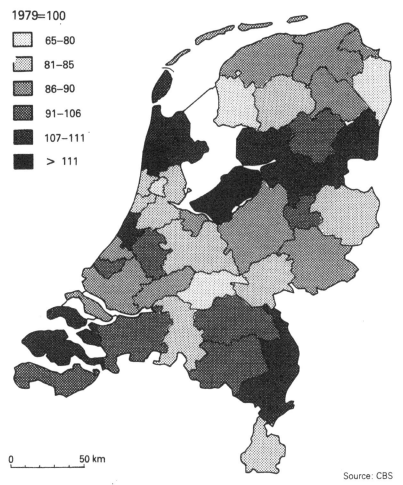

1979=100

░	65–80
▓	81–85
▒	86–90
▓	91–106
■	107–111
■	> 111

0   50 km

Source: CBS

Fig. 4.4   Regional development in manufacturing,
            1979-1985 (employment)

inroads in agricultural processing plants in Oost-Groningen
and in the food sector in the Zaanstreek area, immediately
to the north of Amsterdam. In each instance, these are areas
that stand out on the mental maps of manufacturing imprinted
in the Dutch psyche. Yet there was more at stake than the
decline of traditional manufacturing areas. Nearly all
cities in the western and central parts of the Netherlands
saw manufacturing jobs recede, partly through restructuring
(shipbuilding in Amsterdam and Rotterdam, metalworking in
Utrecht) and partly by the removal of production establishments to the outer rim of De Randstad or sites in the Green
Heart, the non-urbanized area within the confines of De
Randstad.

73

The regional pattern of industry changed after 1973, due in part to the persistent economic recession. Two maps depicting the period between the two energy crises (1973-1979) and the subsequent period (1979-1985) illustrate the de-industrialization of the Netherlands (Fig. 4.3, 4.4).

The regional employment situation after 1973 was disheartening. The traditional manufacturing regions continued to lose ground. The downward trend continued in the large cities as well, to some extent as a result of the ongoing planned overspill to northern Noord-Holland and the new polders (designated growth settlements) and to the Green Heart, the central area of De Randstad. Areas where regional industrialization had previously been successful, like southeast Drenthe, ran into setbacks, Limburg being an exception. The rapid growth of the port areas ceased after the energy crisis of 1973, though manufacturing in Zeeland remained relatively buoyant for a long time.

Industry continued to erode during the first half of the 1980s in places that had by that time become traditional restructuring areas. Yet after 1983, a noteworthy stabilization set in (with Oost-Groningen as the exception). In the areas where growth incentives had been applied the longest, some resilience was noticed; the South and the province of Drenthe are cases in point. In large parts of the West, the disintegration of manufacturing was brought to a halt around 1983. In the North Sea Canal area, however, traditional manufacturing sectors still face formidable problems (shipbuilding, food and kindred products). A new problem area can suddenly appear on the horizon: the riverine area of the province of Gelderland, for example, where hard times have fallen on brickyards, food processing plants, and furniture factories.

Industrial recession may be compensated by employment in the service sector, though this sector certainly does not always absorb all the workers made redundant by manufacturing firms. Industrial disintegration was compensated in the period 1973-1979 by growth in service industries (in absolute terms!) (Table 4.2). From 1979 on, the growth in the tertiary sector reversed in many provinces. Remarkably, the image of a relatively stable position for the incentive areas is confirmed. Moreover, the growth of the tertiary sector in general is strong, at least in the South, even in the first half of the 1980s. The two provinces of Noord-Holland and Zuid-Holland suffered major setbacks, both in manufacturing and in the tertiary sector. The precarious state of manufacturing in the area of Oost-Groningen and the city of Nijmegen has profound repercussions on their respective provinces of Groningen and Gelderland, where manufacturing is now relatively weak. In Groningen, the hoped-for compensation in the form of growth in the tertiary sector

Table 4.2 Employment by province, 1973–1979 and 1979–1984[a]; average annual mutations as percentages, compared to annual mutations in the Netherlands as a whole.

	1973–1979			1979–1984		
	total	industry[b]	services	total	industry[b]	services
Groningen	−1.2	−0.1	−1.6	−0.1	−0.4	−0.2
Friesland	−0.6	0.2	−0.8	−0.3	−0.1	−0.0
Drenthe	−0.3	−0.3	0.6	−0.4	1.4	−0.4
Overijssel	0.1	−0.3	0.6	−0.1	0.2	0.1
Gelderland[c]	0.2	−0.1	0.7	−0.1	−0.6	0.6
Utrecht	0.8	−0.4	1.3	0.5	−0.8	0.5
Noord-Holland	0.0	−0.3	−0.1	−0.5	−0.8	−0.6
Zuid-Holland	−0.2	−0.4	−0.8	0.1	−0.2	−0.4
Zeeland	0.3	2.4	0.3	0.5	1.4	0.6
Noord-Brabant	0.1	0.4	0.9	0.3	0.6	0.8
Limburg	−0.1	0.8	−0.3	0.3	0.9	0.2
The Netherlands	0.3	−2.2	1.4	−1.2	−2.8	−0.2

[a] Labor volume in man-years, estimated on the basis of various CBS statistics.
[b] Including energy.
[c] Including the Southern IJsselmeer Polders.

Source: CPB (CEP 1986)

has not materialized; it has appeared in Gelderland, however (Table 4. 2).

It seems that the old bastions of manufacturing have been virtually dismantled and, in terms of employment, the process of de-industrialization has recently come to a halt. The position of the West is weakening. Previously, the out-migration of firms was responsible for this decline. In the past decade, however, problems of restructuring have become more evident, of which shipbuilding is a case in point. Moreover, the West has failed to attract much-needed foreign investment in manufacturing plants.

The Regional Importance of Industry: Employment

The importance of industry for a region, measured in terms of employment, may be viewed from various perspectives. Industry may provide a substantial number of jobs, but its contribution to the total employment situation can be minor. Yet the reverse may also be the case.

There are still areas that are highly profiled by their industrial activity (Fig. 4. 5). At the top of the list are IJmond, headquarters of Hoogovens, and southeast Noord-Brabant, where Philips reigns. The fenlands in Oost-Groningen, Twente/Achterhoek on the eastern periphery, Zaanstreek north of Amsterdam, and the area around Dordrecht also still bear the mark of their industrial past. The northeastern part of Noord-Brabant may be added to this list. New indus-

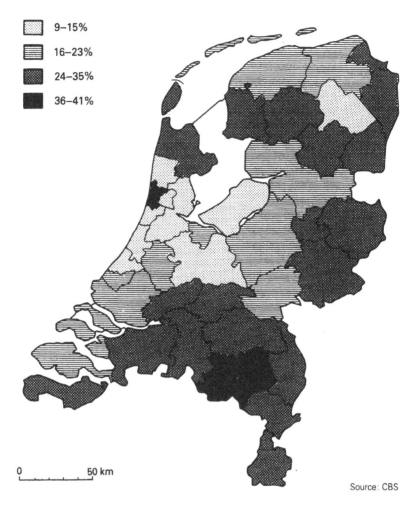

Fig. 4.5   Share of total employment in the manufacturing
           sector, 1982

trial centers are found in the port areas in Zeeland, in
Delfzijl in the province of Groningen, and in Limburg. In
large parts of the country, industry provides more than 30
percent of all jobs for the active labor force. Yet there
are many urban areas in De Randstad where industry provides
only 10 to 15 percent of the jobs; this includes the two
largest office centers, Amsterdam and The Hague.

Does the Netherlands still have industrial areas that
are important in absolute terms? The concentrations near
Rotterdam and Amsterdam are definitely important; Noord-
Brabant is the only other place that stands comparison with
these two areas. In the mid-1980s, almost 100,000 of the

800,000 industrial jobs in the Netherlands are still found in the area of Rijnmond and in the contiguous manufacturing region encompassing Dordrecht and Gorinchem to the east. Nearly 90,000 manufacturing jobs are found in the area of the North Sea Canal, including the Amsterdam agglomeration, IJmond, and Zaanstreek. Noord-Brabant is the paramount Dutch manufacturing province, with more than 160,000 industrial jobs. Eindhoven, Rotterdam and Amsterdam are clearly the top three Dutch manufacturing cities.

The string of cities in Noord-Brabant and the two strongly internationally oriented Dutch industrial zones in Noord- and Zuid-Holland are trailed at a great distance by the hard-hit traditional industrial areas of Twente and Zuid-Limburg, each providing 35,000 to 40,000 industrial jobs. Around 1975, these areas still provided at least 50,000 jobs in industry. A number of less prominent industrial areas with 20,000 to 25,000 manufacturing jobs forms the third level; this encompasses the Groningen fenlands and several urban centers (Utrecht-Amersfoort, Arnhem-Nijmegen, The Hague).

The Demise of Regional Specialization

The industrial map of the Netherlands has become less clearly articulated in the last two decades. Are there still places where a certain branch of industry predominates? This question can be approached by determining the concentration of the following industrial groups:
a.   food and kindred products;
b.   textiles, apparel and footwear;
c.   chemicals (including petroleum);
d.   metallurgical-electric sector (metal, including electrical engineering).

The following definition of a concentration is applied: a concentration should comprise a minimum of 5,000 jobs and, compared to the national proportion, at least 50 percent overrepresentation (location quotient = 1.5).

A map that meets such strict, though arbitrary, conditions (Fig. 4.6) does not leave out many areas that are clearly one-sided. Rijnmond and Zuid-Limburg (DSM) show a concentration of chemical industries, IJmond (Hoogovens) and Eindhoven and its surroundings (Philips) show a concentration of metallurgical industries and electrical engineering firms, respectively. Zaanstreek remains an important center for the food-processing industry; Friesland stands out in dairy products; and northeastern Noord-Brabant is important for meat processing and kindred products in the food, beverages and tobacco sectors. The most striking geographic concentration is in the textiles, apparel and footwear industries. The picture outlined here is what remains of the

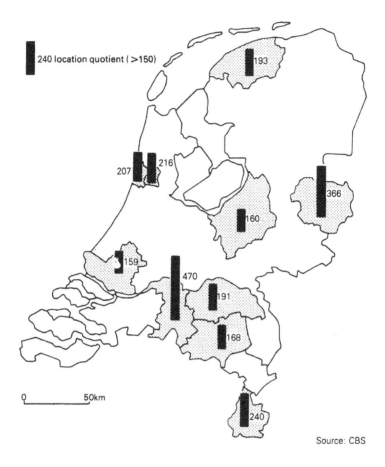

240 location quotient (>150)

193

207 216

160

159

470

191

168

366

0 _____ 50km

240

Source: CBS

Fig. 4.6   Regional concentration of manufacturing, 1982

much more variegated mosaic dating from before the restructuring. In the meantime, many industrial establishments have
disappeared and the locational pattern has become highly
dispersed, particularly in light industrial sectors.

Recent Trends

Which regionally specific strong or weak points have come to
light during the recent economic recovery? A detailed analysis of basic statistical data shows that in a number of
regions the restructuring of weak industrial sectors has run
its course. The apparel industry, though, is still losing
employment in virtually every province; this is also true of
footwear manufacturing in the Langstraat area of Noord-
Brabant. The textile industry presents a different picture,

particularly in Noord-Brabant. The Tilburg wollen industry is holding its own, as is the Noord-Brabant cotton industry. In Twente, cotton is declining slightly. Modernization, with replacement of labor by capital (machines), has already ushered in recovery by increasing the production volume. The loss of jobs in furniture and brick manufacturing is concentrated in Gelderland, which explains the weakened position of the riverine area in this province. A regionally unbalanced impact of restructuring is also felt in shipbuilding. The big shipyards in and around Amsterdam and Rotterdam lost 3,800 jobs between September 1983 and 1985 alone (out of a total loss of 4,390); the northern shipyards and the small docks elsewhere appear to be flexible. Thus, segments of an industrial sector concentrated in a given region prove to have specific characteristics in terms of their market position; this has regionally differentiated repercussions!

It is interesting to see where the growth has recently taken place. It is striking that in the chemical industry (pharmaceutical industry), the paper industry, building materials (cement), metal products, machinery manufacturing, and electrical engineering, Noord-Brabant is a relatively strong grower. The same applies to Limburg, with the exception of the chemical industry (DSM). Zeeland also shows growth (almost) across the board. The industrial center of gravity in the Netherlands is clearly moving south!

How has the restructuring of manufacturing been realized outside the obviously weak industrial sectors? In food and kindred products, Groningen scored low in the recent period of economic recovery. The reorganization in the paper industry seems to have terminated as far as job loss is concerned; the cardboard industry in Oost-Groningen has stabilized at a low level. In heavy industrial sectors, metallurgical activities (Hoogovens) have already seen their worst times, but recently the oil refineries in Rijnmond have cut back ·on jobs. In Zeeland, a recovery in heavy industry is noticeable along the River Scheldt.

4.3 The Regional Importance of Industry: Other Indicators

It is instructive to view sectorial and spatial developments in industry from the perspective of employment, as work is such an important aspect of social life. Yet it should be realized that the use of different criteria can distort the picture. The development of the production volume, influenced by rationalization and automation, can differ from the development of employment. Therefore the outline of the regional situation should be filled in with the details that appear when other indicators are introduced.

Table 4.3 shows the importance of industry within the

Table 4.3   The share of manufacturing in the provincial output (gross
value added) in 1970, 1977 and 1983.

Province	1970	share 1977	1983
Groningen	41.3	63.1	66.9
Friesland	28.3	23.3	14.7
Drenthe	33.5	32.3	38.0
Overijssel	40.3	27.7	21.7
Gelderland	32.3	21.7	15.9
Utrecht	23.8	17.0	12.4
Noord-Holland	30.9	21.8	20.3
Zuid-Holland	30.0	22.5	19.4
Zeeland	35.9	30.7	24.4
Noord-Brabant	44.4	31.1	28.2
Limburg	42.1	29.1	23.4
The Netherlands	33.5	27.0	24.9

Source: Regionaal Statistisch Zakboek 1974, 1982
       Regionaal Economische Jaarcijfers 1983

provincial economy in 1970, 1977 and 1983, on the basis of
the gross value added (the value of the regional product
minus the value of the resources utilized to generate it).
Although the extraction of oil and gas introduces distor-
tions, particularly for Groningen, the overall picture is
clear. The industrial provinces are found in the South and
in the North, both in 1970 and in 1983. Friesland is an
exception in the North. This is largely due to the contribution of the agricultural sector. In Friesland, this contribution (10 percent in 1983) is considerably greater than in
other provinces. In terms of gross value added, the provinces of Utrecht and Gelderland have a weak industrial orientation.

A somewhat comparable picture can be drawn on the basis
of the contribution of manufacturing to export. The highest
contributions are found in the northern (except Friesland)
and southern provinces (Fig. 4.7). Utrecht, Noord-Holland,
Zuid-Holland and Gelderland all show much lower values.
Utrecht has the highest contribution in the wholesale branch
(23.6 percent). In Gelderland, besides agriculture (6.5
percent), wholesaling (12.3 percent) is also well represented. Noord-Holland and Zuid-Holland have a high contribution from the seaport and airport sector.

The analysis at the level of the province can be supplemented with data on the COROP regions. In Fig. 4.8 the
investments in fixed industrial assets are shown per COROP
region, in relation to the size of the population. Comparison with the provincial data demonstrates that especially
the industrial area of Delfzijl contributes to the relatively high score for the province of Groningen. Rijnmond
fills this role for Zuid-Holland, as does Zeeuws-Vlaanderen

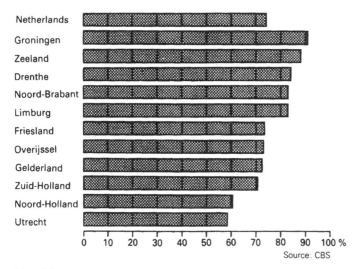

Fig. 4.7   Share of provincial export contributed by the
manufacturing sector, 1983

for Zeeland. It is also noteworthy that, compared with
Rijnmond, there is relatively little propensity to invest in
the three remaining metropolitan areas of Greater Amsterdam,
the agglomeration of The Hague and the province of Utrecht.
This once again illustrates the difference in economic specialization between these areas. Noord-Holland scored relatively low despite high per capita investment in the traditional industrial areas of Zaanstreek and IJmond.

The short-term dynamics can be deduced from Fig. 4.9.
Relative to the average investment in industry in the period
1978-1980, growth occurred in three areas: southwest Drenthe, southwest Overijssel and in Westland/Delft, the region
between The Hague and Rotterdam. However, when expressed as
per capita investment, these areas attracted no more than an
average level of input. In addition, the investments in the
traditional manufacturing area of Zeeuws-Vlaanderen increased substantially. It is also noteworthy that most areas
with a high per capita level of investment only experience
an average rate of growth (Rijnmond, IJmond) or even lag
behind the national pace of growth (the southeast of the
province of Noord-Brabant, Noord-Limburg, Zuid-Limburg).
Since most areas with a low per capita level of investment
show a growth rate somewhat higher than the national average, this would indicate a flattening out of the regional
differences. Comparable with earlier data on the provinces,
Fig. 4.10 depicts the proportion of manufacturing in the

81

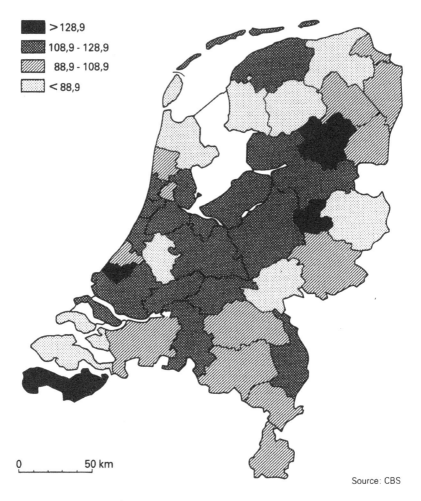

Fig. 4.8   Average investment in fixed assets in the
            manufacturing sector, 1981–1984 (related to
            size of population on 1 January 1983)

regional export package. The areas most strongly oriented
toward the export of manufactures are the traditional industrial areas: Delfzijl, southeast Drenthe, IJmond, Zaanstreek, Zeeuws-Vlaanderen, Eindhoven and Zuid-Limburg. The
figure also highlights the strong place of manufacturing in
the export from the entire Northeast and the great importance of industrial export for Rijnmond in comparison to the
other large agglomerations. The areas where the industrial
contribution to export is relatively small are occasionally
strongly oriented toward agriculture (Westland: 43.8 percent!), mineral extraction (the northern part of Noord-

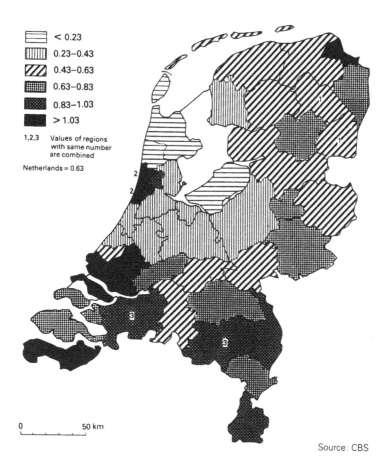

☰	< 0.23
⦀	0.23–0.43
⧄	0.43–0.63
⊞	0.63–0.83
▦	0.83–1.03
■	> 1.03

1,2,3   Values of regions
        with same number
        are combined

Netherlands = 0.63

0 ⌊_____⌋ 50 km

Source: CBS

Fig. 4.9   Average investment in fixed assets in the
            manufacturing sector, 1978/80 – 1981/84
            (1978/80 = 100)

Holland: 54.4 percent), or services (Greater Amsterdam and
the agglomeration of The Hague).

The picture is completed with data on the proportion of
industry in the gross value added per COROP region (Fig.
4.11). In broad terms, the pattern has already become com-
monplace. The preeminent manufacturing areas are now Delfzijl, southeast Drenthe, IJmond, and Eindhoven. The position
of the rest of Groningen is tied to the exploitation of
natural gas. The areas that are only slightly oriented
toward industry display a wide variety of specializations.
One example is that in southwest Friesland and in Westland,
the contribution of the agricultural sector is greater than
that of manufacturing. The region of De Veluwe, a very
attractive residential area adjoining De Randstad to the

83

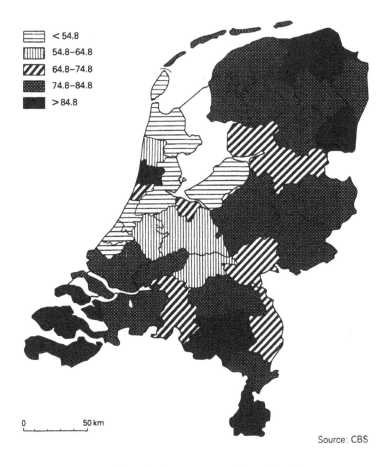

☰	< 54.8
⫿⫿⫿	54.8–64.8
⧄	64.8–74.8
▦	74.8–84.8
■	> 84.8

0 _____ 50 km

Source: CBS

Fig. 4.10   Share of regional export contributed by the
            manufacturing sector, 1983

east, registers the second-highest contribution of government (22. 8 percent) to its economy after The Hague (23. 2
percent). The contribution of the government sector to the
local economies of greater Amsterdam and Greater Rijnmond
only reached 12. 7 percent and 9. 3 percent, respectively.

4.4 Summary

This chapter utilizes employment statistics to show that
until 1963, clear regional-economic specializations could be
identified in the Netherlands. Subsequently the picture
became fragmented. Many areas gradually lost the traditional
industrial character peculiar to their development. At the
same time new manufacturing areas emerged.

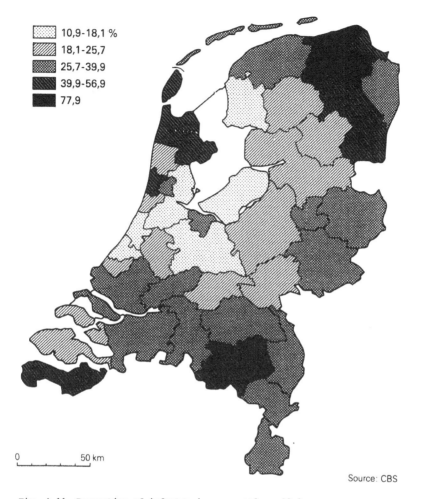

Legend:
- 10,9-18,1 %
- 18,1-25,7
- 25,7-39,9
- 39,9-56,9
- 77,9

0 _____ 50 km

Source: CBS

Fig. 4.11 Proportion of industry in gross value added per
region

    This image of fragmentation of regional specializations
was not only confirmed but even reinforced by studying other
indicators, namely, investments in fixed assets, the propor-
tional contribution of industry to exports, and gross regional product. In terms of the proportion of the gross regional product generated by industry, the most industrial
regions are Delfzijl (Groningen), southeast Drenthe, IJmond
and Eindhoven. In the next category, some traditional industrial areas are found alongside some new ones: Twente, Zaanstreek, Rijnmond (including Dordrecht and its surroundings),
in addition to regions like Zeeuws-Vlaanderen and the western part of Noord-Brabant.

5 Industrialization, Policy and the Region

The changes in the regional distribution of industrial employment, which were outlined in the preceding chapter, are partly related to the regional policy implemented by the national government. Until recently, regional policy was essentially industrialization policy. The first steps toward forging a regional policy were taken shortly after the Second World War during the period of economic reconstruction.

5.1 Regional Industrial Policy in the Making

In Chapter 2 we showed that the industrialization offensive that was begun after 1945 had been given national priority. Much had been done to gain the active support of governmental departments and industrial sectors, represented by employers' organizations and trade unions. In addition, the concept 'industrial climate' was often used in the industrialization memoranda. The 'industrial climate' is similar to the concept of production environment and embodies the external conditions for production that do not appear on the company books, such as infrastructure (roads, public utilities), industrial estates, training centers, housing, etc. Good external conditions were scarce at the time, particularly in certain parts of the country. Still, this was not considered reason enough to implement a specifically regional policy. In 1951, moreover, the registered labor reserve for men fell below 3 percent, the so-called lower limit of Beveridge. Thus, there was officially no structural or conjunctural unemployment.

The reason to implement a regional policy, despite this low figure, was that high unemployment persisted in several regions. The fenland districts in the northeast of the country, were confronted with structural unemployment due to the phasing-out of peat cutting (8 percent in the eastern part of the province of Groningen, 13.8 percent in the adjacent part of southeastern Drenthe). In parts of Noord-Brabant (southwest and northeast) the growth of employment lagged behind the rapid increase in the size of the labor force (Vanhove 1962: 9).

Regional development plans have been drafted since 1946. The issue of structural unemployment was seen as the precondition for a regional industrialization policy. The first initiatives for a regional approach to increasing well-being were made as long ago as 1932, with the establishment of the Limburg Economisch-Technologisch Instituut, a government-sponsored economic consulting agency. This was followed in 1935 by the NV Industriebank for Limburg, a quasi-governmental organization for financial participation. Such initiatives, however, were strongly denounced by spokesmen of the national government, which disapproved of regional financing institutions (De Rooij 1979). Numerous reports were produced at that time, presenting research results on regional sources of prosperity. These were mostly written by human geographers, including Van Vuuren, Ter Veen and Boerman (Stolzenberg 1984). The government-appointed commission for industrialization produced a report in 1950 propounding the national dispersal of industrialization by regional concentration. In this report, the scope of regional industrialization was much wider than a mere tactic to combat structural unemployment. In line with the report on the distribution of the Dutch population (Kortlandt 1949), a range of arguments was presented to maintain the dispersal of the population. These included the social costs of migration, the growth of the supply of labor, the strategic vulnerability of concentration in time of war and, finally, the high cost of construction in the West. This dual track (unemployment and dispersal arguments) characterized all ensuing regional industrialization policy.

5.2 Regional Industrialization Policy:
From Fighting Unemployment to Population Dispersal

As the unemployment rate declined in the nine development areas identified in 1952, the regional industrialization policy appeared to have accomplished its goal. Yet at the same time, the importance attached to this regional policy in subsequent industrialization memoranda kept growing! An increasing portion of the country was covered by this policy and the policy instruments were extended significantly. This paradox begs an explanation. First, however, we elucidate the initial stage of the regional industrialization policy. The explanation of the policy reorientation follows. Finally, attention is given to the explanation of the changes in the geographical pattern of industry in the Netherlands.

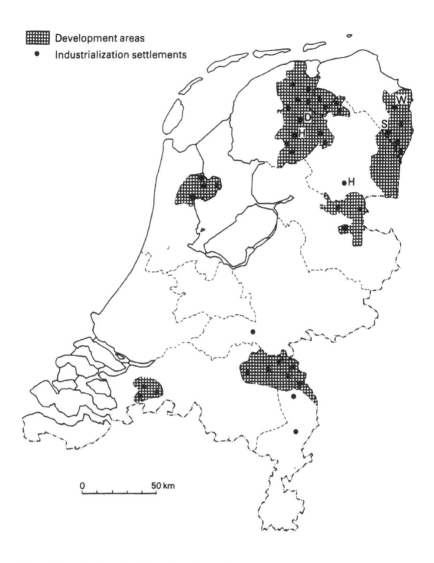

Fig. 5.1   Regional policy in the period 1952–1958
           (stage 1)

Fighting Structural Unemployment in Development Areas

In 1952, after a first stage of successful industrialization, legislation was enacted to permit the promotion of
industrialization in eight 'development areas'. Following
the example of southeast Drenthe, given the status of a
development area in 1951, this law formed a basis on which
to combat the structural unemployment that had been cropping

up there (see Section 2. 1). Thirty-eight industrialization centers were designated in these areas and four elsewhere; virtually all of these centers had a population ranging from 15,000 to 40,000 (Fig. 5. 1). Each town was expected to have access to the means to accommodate industrial establishments (industrial estates, housing, training centers, etc.) and had to be made accessible (plans for road construction). Firms that located in these centers received a premium for each unemployed person they hired. In addition, migration abroad or to the West of the Netherlands was promoted by reimbursing out-migrants for part of the expense of moving. From 1952 to 1958, 3,600 unemployed married men took advantage of this opportunity, of whom two-thirds came from the North and half chose the West as their destination (Klaassen & Drewe 1973).

One problem was how to include older persons and individuals for whom job placement is difficult due to social reasons. The regional industrialization policy was therefore extended to include social work in order to guide the rural population through the transition from agriculture to industry. This transition entailed a different work tempo and discipline. It was feared that the population would become anti-social, secularized and irrational, since the areas designated for development were deprived in many ways. Numerous social and cultural, even sanitary facilities were lacking (Van Doorn 1960). The improvement of the social climate was to benefit the industrial climate. It was supposed to facilitate a smooth adaptation to the norms and values of industrial society.

However, the designated development areas differed sharply in the nature of their economic base and their population dynamics. In the northern development areas, the central problem was the expulsion of labor from agriculture and peat-cutting (southeast Drenthe). In the South, the vigorous growth in population played a dominant role. The policy instruments were also aimed at combatting unemployment. The more indirect measures were expected to achieve an even greater effect in the long run (industrial estates, housing, services, roads).

The results of the first stage of the regional industrialization policy (1950-1957), which coincided with the second stage of industrialization and is documented in the sixth memorandum on industrialization policy, were remarkable in two respects:

a. The development areas showed an increase of 15,200 industrial jobs, a growth of no less than 53 percent. In most areas, 1,500 to 2,500 new jobs were created (and 2,738 in southeast Drenthe), though a sizeable proportion of these jobs were in large Dutch firms. Philips, for example, provided 1,726 of the 2,362 new jobs in the eastern part of the province of Friesland (on the contribution of Philips, see George 1961, Fischer 1980, Tromp 1958).

b.   Industry expanded in other regions outside the Western provinces as well (by 69,700 jobs). This trend resulted in a decline of the share of industrial employment in the West from 44.1 percent to 42.1 percent. An autonomous trend toward deconcentration occurred.

The regional industrialization policy was not embraced by everyone. In a study carried out by the research bureau of the Labor Party, provincial departments were criticized for their 'extremely one-sided view of how to absorb the oversupply of labor locally and the lack of concern for the promotion of geographic mobility' (Steigenga et al. 1955). They were cautioned that the presence of cheap labor should not be overestimated as a factor to attract new industrial establishments, since this advantage might quickly disappear. The report agreed with the argument of Hofstee (1950) that regional differences in population pressure would not have a great impact on the regional development of industry. Earlier, Steigenga (1949) and Vermooten (1949) had expressed conflicting views on regional industrialization. The former defended the standpoint of 'people to jobs' (meaning to the West) on economic-geographical grounds, and the latter advocated 'jobs to people' on social grounds. The principle of dispersal would emphatically support the latter point of view.

Dispersal Policy

Regional dispersal policy has occupied an important place in the industrialization memoranda published since 1958. Although the fear of insufficient employment in the development areas was to remain in the background, attention was directed primarily toward the spatial dispersal of industry. The social goals were allocated a place in the strategy, but they came after the economic and closely related spatial targets. The government was confronted with a problem in the western part of the country: the West demonstrated a high growth potential but had to deal with a housing shortage, traffic congestion, incursions into the landscape and arable land by urban expansion, shortage of recreational space, etc. Such problems were extensively treated in memoranda ('The West and the Rest of the Netherlands' (1956) and 'The Development of the West of the Country' (1958)). The selective distribution of industry would be desirable from a planning perspective. At the same time, it should be avoided that growing industries, especially those which require locations on deep water or in a large city, would be confronted with serious shortages of labor. In fact, a

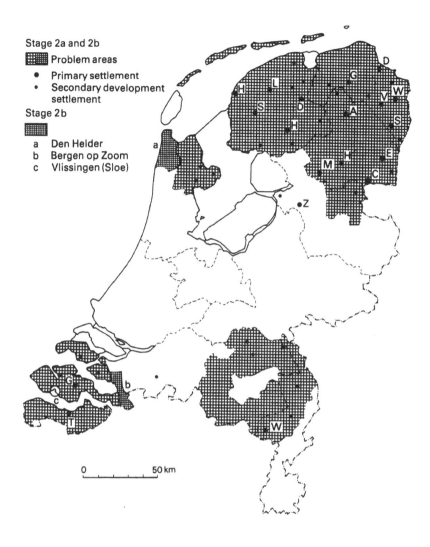

Stage 2a and 2b

Problem areas

• Primary settlement
• Secondary development
  settlement

Stage 2b

a   Den Helder
b   Bergen op Zoom
c   Vlissingen (Sloe)

0 _____ 50 km

Fig. 5.2   Regional policy in the period 1958-1968
            (stage 2a and 2b)

three-pronged regional policy took shape: exploitation of
the economic-geographical advantages of the West, in part by
spatial dispersal of jobs not tied to the West, and supplemented by attention to a number of development areas to
foster social justice. Nonetheless, it would be a long time
before regional policy was linked to regional potentials.
The dispersal policy was based on incentives, not restrictions (no system of permits to allow establishment of industry in the West was instituted). Wherever opportunities for

91

economic development appeared, outside the West, they should be seized.

Expectations regarding migration and unemployment played a viable role in the selection of the new development areas. Furthermore, the areas had to be larger than the 'old' development areas in order to offer ample alternatives for relocation and to allow a comprehensive development of these parts of the country. The combined surface of the development areas covered almost 40 percent of the Dutch territory (the previously designated development areas covered 17 percent). One-fifth of the Dutch population lived within these development areas (Fig. 5.2).

There was increasing difference of opinion between the Ministry of Economic Affairs and several provinces on whether areas where jobs were scarce and commuting was rising steeply should be designated as development areas (Van Hoogstraten 1983: 118-121). Two areas demonstrated not only a low level of unemployment but also a high degree of migration or commuting: the northern part of Noord-Holland, the far north of the province, which was the recruitment shed of Hoogovens iron and steel works; and the western part of the province of Noord-Brabant, where one out of every four wage-earners worked in the industrial estates and port areas of Rotterdam and Dordrecht. These areas were excluded from the regional policy. In fact, the government did not want Hoogovens and the port of Rotterdam, which were spearheads of the Dutch economy, to have to cope with a shortage of labor (Ter Hoeven 1963). A specific form of dispersal policy was conceived whereby functions that depend on deep water would be developed along the Westerschelde Estuary and in the North near Delfzijl.

The Ministry of Economic Affairs also disagreed with the provinces on which places should be considered for the status of development centers. The provinces advocated designating regional centers of as many service areas as possible. The Ministry, however, preferred places with strong development potential and/or locations where large firms would invest if the industrial climate were improved, for example Philips in Drachten and Stadskanaal (see Van Hoogstraten 1983: 316).

The new stage of regional policy went into effect in 1959. Of the 42 development centers, only 19 were retained in the new program (Fig. 5.2). A coarser pattern of centers was drawn up; better traffic access and new vehicles (the light motorbike!) made this feasible. The goal was to avoid fragmentation of the industrial estates and facilities. The enlargement of the total program area in the latest stage, however, resulted in the designation of 44 development centers in total (of which 25 primary and 19 secondary). The policy instruments were intensified considerably. Subsidies on job placement, including employment of individuals who were not out of work, fitted in with the old policy. Bonuses on buildings according to floorspace and discounts (50 percent) on purchase of land zoned for industry were out-growths of the distribution goal that focused primarily on large projects. The infrastructure programs were consistent with this goal. Migration to development centers was facili-tated by reimbursement of the expense of moving. Foreign emigration and moving to the West were no longer promoted.

The total amount of funds for regional-economic policy was only 2 to 3 percent of the budget of the Ministry of Economic Affairs in the early 1950s; in 1963, this was more than 10 percent (Bartels & Van Duijn 1981: 126). It is striking that most of the funds for infrastructure ended up in the North (92 percent), predominantly in the province of Groningen. The North was supposed to be pulled closer to the West in order for the dispersal policy to succeed. The dispersal policy was not only manifest in the regional industrialization policy. The Agency for the National Plan had also emphasized dispersal since the publication of memoranda such as 'The West and the Rest of the Netherlands' (1956) and 'The Development of the West of the Country' (1958). Some plans failed, including an initiative to move the bureaucratic apparatus of government from The Hague to Apeldoorn and the designation of Apeldoorn as the country's second-ranking administrative center. But plans to move government departments to Apeldoorn, Arnhem and Amersfoort became more concrete around 1960 (De Smidt 1985b). The fact that the regional industrialization policy had not yet been turned into a general regional-economic policy in which the service sector played a role explains the failure to disperse more office activities.

In his evaluation of the effect of the regional policy, Vanhove (1962) was able to identify two significant variables for regional growth of employment at the level of the economic-geographic area: the pool of excess labor and the wage level. Incidentally, the regional wage differential in manufacturing was not great: 10 percent below the national average for unskilled labor in Drenthe; 5 percent above average in the provinces of Noord-Holland and Zuid-Holland. Moreover, Vanhove had evidence that transport costs, certainly in the North, were an obstacle to decon-centration. Other factors such as sales markets, agglomer-ation economies, construction costs, land prices and bonuses did not influence deconcentration in either a positive or negative sense. It is striking that Vanhove gave a more positive assessment of the southern development areas than the northern ones (with the exception of eastern Friesland, dominated by Philips). The areas in Groningen did not meet expectations, nor did the southeastern part of Drenthe, where industry had to cope with a pool of excess labor that proved to be only partly suited to industrial work (Vanhove 1962: 103-4).

Analysis with the shift/share method for the period 1950-1963 revealed an autonomous deconcentration trend (Wever 1971). Despite a good mix in the composition of industry, the western provinces were not able to cope due to disadvantages in the production environment (supply of labor, congestion, etc.). The opposite was true for Drenthe, Friesland and Limburg in this period. Noord-Brabant also profited from the deconcentration, but structural problems had already cropped up in the central part of Noord-Brabant (Tilburg and Langstraat). Groningen and Overijssel - just as Vanhove observed - did not benefit from the deconcentration and struggled with structural problems. Zeeland had a very limited industrial perspective to offer. These conclusions at the provincial level do not contradict the fact that this method of analysis revealed the same problems at the regional level as those found with the method used by Vanhove.

The results of regional industrialization, whether brought about by policy or not, can be simply illustrated. The industrial employment in the West of the Netherlands increased between 1953 and 1963 by 38,000 jobs (9 percent), and outside this area by 147,000 (28 percent). The number of jobs in new industrial establishments was more than 22,000 in the West, and elsewhere more than 62,000. Actually, on the basis of the regional shares in 1953, the West should have been able to provide 37,000 jobs in new establishments; nearly 15,000 jobs thus went to a different region than expected. It is equally important to recognize that, outside the West, the effect of expansions (minus closures) was relatively much stronger than the effect of new establishments, despite the dominance of some structurally weak industrial sectors in the eastern part of Groningen, in

During the peak of economic growth in the 1960s, the trend toward autonomous dispersal was buttressed by the acute shortage of labor, particularly in the West. In this part of the country, the pressure of economic activities increased further because of the immense growth of the office sector in the large cities and the expansion of the port-related industrial activities (De Smidt 1973). The seaports, primarily Rotterdam, felt the impulses of an increasing orientation of Western Europe toward deriving its raw materials from overseas, which invited the establishment of metallurgical and petrochemical industries (Wever 1974). The overspill of seaport-dependent industry to Zeeland and the northern city of Delfzijl was predicted in the seaport memorandum (1966) and was supported by regional policy (Fig. 5. 2). A pipeline network was installed between Rotterdam and

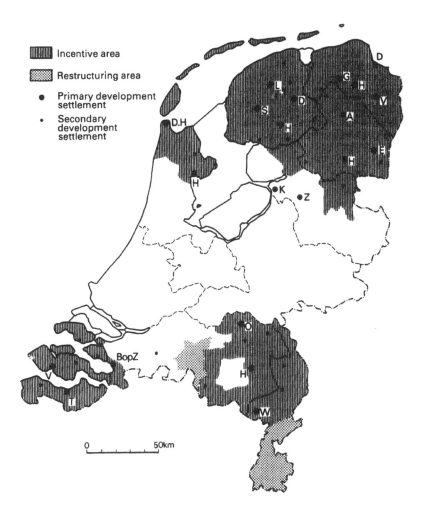

Fig. 5.3   Regional policy in the period 1969-1972
            (stage 3)

Antwerp, which ran through the Schelde Estuary harbors in
Zeeland. The province of Zeeland, which was incorporated in
the incentive policy in 1965, did not need this policy any
more around 1971 and, along with former incentive areas in
Noord-Brabant, ceased to receive government support under
this program.

    The incentive policy, intended to gradually help indus-
trialize a region by improving its industrial climate, was

augmented in the second half of the 1960s to include another form of regional policy: the restructuring policy (Fig. 5.3). This policy focused on urban industrial areas that were highly dependent on one industry group and were subject to an invasive restructuring process involving the loss of thousands of jobs in a short time. In these areas, such as Zuid-Limburg (coalmine closures), Twente, Tilburg and Helmond (slump in the textile industry), an industrial climate was already formed. New sources of income had to be found quickly. In this light, it is understandable that a more potent policy instrument was required, as in other countries. The investment bonus program (IPR) was enforced in 1967 to provide a subsidy on capital. It was better suited to create a new industrial pattern with higher investments per worker. The job subsidy faded gradually into the background.

5.3 Regional Policy in a Time of Recession and Selective Growth

Around 1972, after two decades of regional policy, there seemed to be little need to keep providing such a large part of the Netherlands with employment incentives. A long-term policy for the North and for Zuid-Limburg could be fitted into the much wider framework of a policy for the dispersal of population and employment (Fig. 5.4). The pressure exerted on space in De Randstad by the port-related industrialization and the office boom would thereby not be increased. For the first time, incentives in the peripheral areas would have to coincide with restricting measures in the West. This combination of goals and instruments had already been incorporated in the policy in other countries. Experience there showed that a ceiling on investment in metropolitan core areas did not lead to dispersal over long distances. It did lead to fringe growth, in fact to growth just beyond the limits of the metropolitan area to which the mitigating measures applied. This area is commonly described as the intermediate zone.

The implantation of growth poles in peripheral areas was the cornerstone of the regional policy in the second half of the 1960s until well into the 1970s in numerous industrial countries. The growth pole consists of a firm or a combination of firms that, by input and output relations, could induce more firms to establish in the same area (technical polarization) or, by the way they spend their income, could provide more jobs for local or regional industry (Vanneste 1967). Moreover, such a growth pole could attract firms that are not functionally tied to this complex (psy-

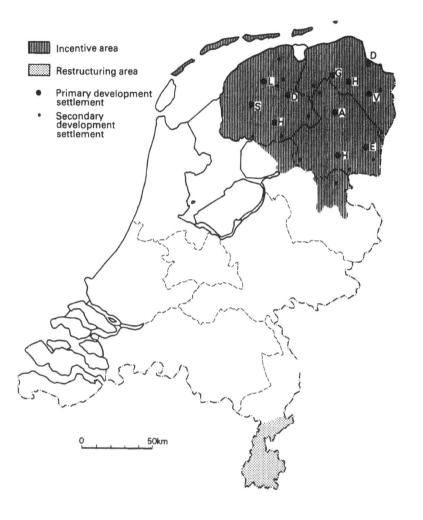

Incentive area

Restructuring area

● Primary development settlement

● Secondary development settlement

0   50km

Fig. 5.4   Regional policy in the period 1972–1978
(stage 4)

chological polarization). Eventually, a concentration of industrial activity, and thus of population, could counterbalance the existing urban areas and relieve the pressure on the metropolitan centers (geographical polarization). One of the opportunities for peripheral areas was to take advantage of the capital-intensive incentives; these benefit the metallurgical and petrochemical primary industries which make up most of the engines of growth. In addition, the peripheral areas were supposed to take advantage of the

97

locational differentiation process within the social division of labor, which, according to Goddard (1975), consists of 'orientation, planning and programmed' contacts. The orientation contacts are most strongly tied to a metropolitan environment, while the programmed contacts are the least dependent on the city. The peripheral areas should certainly aim to attract the less-specialized programmed activities. This was not necessarily going to be successful since the competition with countries with low wages would clearly set limits on this endeavor. It was therefore important also to stimulate more highly specialized activities; the North, given its increasingly sophisticated labor supply, was considered to be a successful candidate for such activities. As a complement to the industrialization policy and its focus on primary industry, the stimulation of the service sector in peripheral areas was included as an element of regional policy. This was also the motivation to stimulate the decentralization of offices of government agencies (De Smidt 1985b).

Regional policy thus became less exclusively concerned with regional industrialization and became increasingly involved in the spatial dispersal policy and labor market policy. Eventually, following the memorandum on selective growth (Economische Struktuurnota 1976), the frame of reference for regional policy was widened to include other facets such as environmental protection, resource and energy conservation and international division of labor. This made regional policy in the 1970s much more complicated than in the 1950s and 1960s. Above all, it became difficult to implement a regional policy around 1972 when the recession set in. In this long-lasting recession, the erosion of industry (discussed in Chapter 3) and the limited growth in the service sector left little leeway for a regional redistribution of resources.

The goals of the regional policy in the 1970s should be seen against the background of the processes outlined above. The policy for the North and for Zuid-Limburg was continued, with an emphasis on the creation of growth poles and metropolitan production environments. In this task, attention was focused on continuing the industrialization process and also on activities in the service sector (including the deconcentration of national government agencies). The other previous stimulation areas (particularly in Noord-Brabant) were expected to develop further on their own. A policy of setting limits to growth was enforced in the West, which could also benefit the peripheral areas (Fig. 5.5).

After 1977, the guiding light for regional policy was the mitigation of socioeconomic imbalance among the regions in the Netherlands. In the short run, however, the rapid

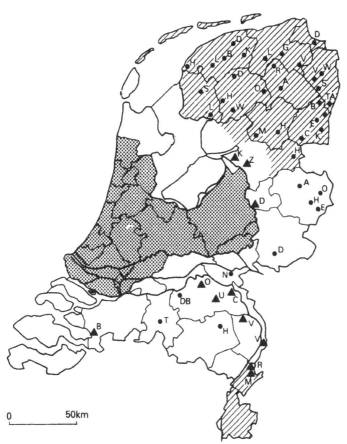

0 _____ 50km

▨ Northern incentive area and restructuring area of Zuid-Limburg

▨ Area for selective investment regulation

●   Centers with investment premium (25 percent of greenfield and
    extension investments) for manufacturing and sevices establishments

▲   Centers with investment premium (15 percent of greenfield and
    extension investments) for manufacturing establishments

Fig. 5.5   Regional policy in the period 1978–1981
          (stage 5)

growth of unemployment demanded a forceful policy.

    The instruments of regional policy were not altered in
essence as far as industry was concerned. In response to the
recession, the standard package, which included the investment bonus program, accelerated depreciation and subsidies
on infrastructure, was bolstered by enacting measures to
support individual firms and by implementing employment programs. These two instruments, however, should be seen as

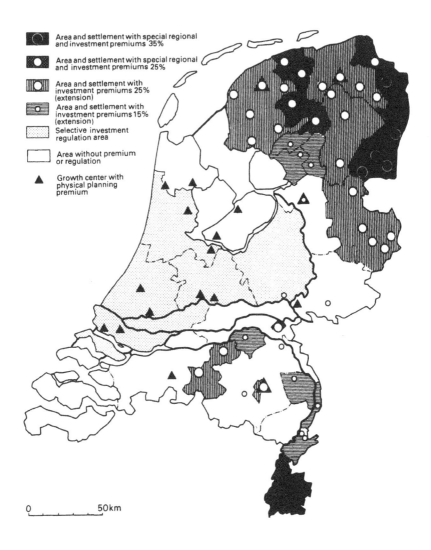

Area and settlement with special regional
and investment premiums 35%

Area and settlement with special regional
and investment premiums 25%

Area and settlement with
investment premiums 25%
(extension)

Area and settlement with
investment premiums 15%
(extension)

Selective investment
regulation area

Area without premium
or regulation

▲ Growth center with
physical planning
premium

0   50km

Fig. 5.6   Regional policy in the period 1982-1985
(stage 6)

general measures rather than as parts of a regional policy.

The first disincentive in the area of regional policy
was introduced in 1972 and was put in operation on 1 October
1975 in the West and in the central part of the Netherlands.
This was the selective investment act (SIR). In the area
circumscribed in Fig. 5.5, permission could be denied for
the construction of a new establishment or the expansion of

100

an existing one on the grounds of its contribution to the concentration of activities and population, (one-sided) economic structure, and an (unbalanced) labor market. The request could be denied, that is, if the compulsory reporting would indeed lead to the initiation of an official permit procedure. This permit procedure was instituted in the regional jurisdiction of Rijnmond. Elsewhere, the authorities did not enact such a procedure, but in individual cases, it could be required by the Minister of Economic Affairs (as happened, for example, in a case involving a development in the wooded hills to the east of the city of Utrecht). Its application shows that although this instrument had a preventive effect, it did not generate the desired structure (i. e. diversification) in Rijnmond.

The existing instruments were further intensified. The investment bonus act remained in force. The new (1978) act for investment accounts (WIR) offered the government a range of regionally differentiated support measures, including the so-called special regional bonus (BRT) (Fig. 5.5).

In addition to more powerful instruments, the regional socio-economic policy for the North and for Limburg was embedded in a comprehensive regional development plan, in which spatial and socio-cultural policy also received attention. The comprehensive structure plan for the North (ISP 1979) and the perspective memorandum for Limburg (PNL 1978) contain five-year plans based on a wide approach to regional development (De Smidt 1985c). The entrenchment of the economic recession around 1980, however, put many good intentions to the test. In the period 1981 through 1985, these two regional plans were the key instruments of the national policy and the BRT was primarily applied in these areas (Fig. 5.6). The selective investment program was initially continued, although the growth centers close to the large cities were to receive a spatial planning bonus (ROT) for receiving establishments from the donor cities.

5.4 Recent Trends in Regional Policy

The economic recession that appeared around 1972 has since then determined the perception of regional policy and its contents to a large extent. The recession caused a reversal in the trend of economic growth. Decline generally implies contraction and closure of firms; and the spatial distribution of these activities is not amenable to governmental steering. The most important policy instrument in regional policy, the IPR (investment bonus program) is in essence a growth incentive. The economic recession also caused unemployment to rise almost everywhere. In such a situation, the policy to disperse economic activities is naturally less pertinent than in a period when a tight labor market in one area coincides with unemployment elsewhere. Specifically, the traditional donor area in the dispersal policy, the large cities composing De Randstad, witnessed an alarming rise in unemployment after 1972. In that situation, it was politically and economically unreasonable to advocate further dispersal. Therefore, the deconcentration of governmental departments was reconsidered, as was the SIR. In fact, this investment deterrent program for the West had hardly any impact on the peripheral areas. In 1983 it was put on hold and in 1985 it was abolished.

The ubiquitous rise in unemployment deprived the peripheral areas of their strongest bargaining chip in the period 1958-1972: the presence of a labor force that was scarce elsewhere. In addition, local and regional governments, including those in the West, made greater efforts to retain their own industry. Even industry that had previously been considered unsuitable for the metropolitan environment found itself in the good graces of the municipal authorities. Because this undermined the foundations of the dispersal policy as it was enforced until then, the perspectives for the peripheral areas ebbed. The spatial planning policy was adapted. The increasing unemployment in the large cities made it less opportune to implement deconcentration policy there. The fear that the metropolitan areas would lose their economic base led to a reorientation toward the large cities. The idea dating from the 1960s that the cities were becoming unlivable faded rapidly.

A third development with a negative impact on the peripheral areas was the shift from the regional to the national economic problem. It became increasingly accepted that the central task of the government should be to stimulate national economic growth. The unacceptably high rate of unemployment in nearly all regions means that every chance to create new jobs must be seized, regardless of where it is located. National spearheads, such as the port of Rotterdam and Schiphol airport, which had been subject to restraint in the 1960s, were again considered to be national growth poles and they then received the attention they had deserved previously. On the one hand, this illustrates a weakening of the position of the regional policy and a strengthening of the so-called generic policy. On the other hand, the recession pushed the help for weak regions into the background in favor of the strong regions. In this connection, a distinction is made between spatially discriminating incentive policy and the non-spatial discriminating development policy. The previously mentioned innovation policy (such as support for bio-technology) and special target area policy (support for the Gateway-to- Europe function) fit into the development policy.

There was also an essential shift in regional policy.

In the period of growth, the policy, including the recruitment of foreign firms, was highly centralistic, whereby all
activities were coordinated in The Hague. Now that the
policy depends on generating new initiatives, the regions
have come to the forefront. The 'organizational capacity' of
the region is now of prime concern. This concept does not
have a spatially discriminatory effect. Each region can
develop initiatives. To finance the initiatives, however,
some areas only have recourse to generic programs. In other
areas, the regional incentive policy offers additional
opportunities. Regional development agencies have taken
advantage of these options. The oldest of these, the Northern Development Association (NOM), was established in 1974.
It is financed by the national government and its charter
allows it to participate in firms. The NOM is the actual
owner of a number of firms in the North. In Zuid-Limburg,
the LIOF, the Limburg Institute for Development and Finance,
fulfils a similar function. Because of the area in which
they operate, these two regional development agencies (ROMs)
received the most support from the national government. The
development agencies of Overijssel (OOM), Gelderland (GOM)
and Noord-Brabant (BOM) work with more modest budgets and
rely more on regional financing sources.

The recent shift in orientation within the regional
policy, as well as the recession and the concomitant
budgetary austerity, have blunted regional policy instruments. Earlier, the special regional bonus (BRT) had been
retracted from the non-spatially discriminating WIR, partly
because the EC so demanded. The current slate of bonuses
(1986-88) (Fig. 5.7) shows that only the - regionally
differentiated - IPR remains. The number of supported areas
is declining; hence, parts of the province of Drenthe are no
longer in the picture.

In 1988 the above-mentioned duality in regional policy
showed up even more clearly in the so-called re-standardiz-
ation memorandum that covers the period from the present to
1990. With a constant budget, it calls for a transition from
investment support for disadvantaged regions (IPR) to pro-
ject-oriented support for new technologies and other 'structure-bolstering' measures. These are described below. Although the principle of equity is not discarded, funds are
transferred to projects that support the goal of efficiency,
that is, that are focused on bolstering macro-economic
growth. The IPR areas are limited in number to those regions
where the unemployment - despite national economic recovery - is still extremely high. In the South, only Zuid-Limburg and Helmond are left, both typical restructuring areas
(cf. Figs. 5.6 and 5.7). Also, the bonus system of the IPR
has been considerably simplified. Now only one bonus percentage (25 percent) is awarded upon establishment and expan-

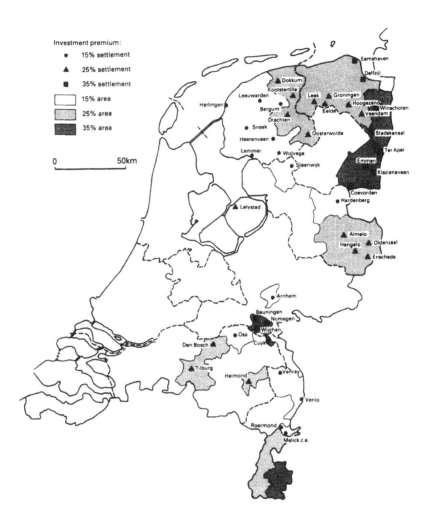

Fig. 5.7   Regional policy for the period 1986–1990
            (stage 7)

sion within 5 years (thereafter, the bonus drops to 15
percent.)

The existing regional programs for the North (ISP),
(Zuid-)Limburg (PNL), Twente and Nijmegen are being
maintained, yet these programs must prepare for an internal
shift from policy focused on generating the preconditions
for investment (provision of infrastructure, industrial

estates) to support for structure-bolstering projects that lead to a higher contribution by the regional economy to the macro-economic growth (efficiency goal). The latter goal, however, is not only applicable to disadvantaged areas, but it covers all seven provinces outside the West. The grounds for circumscribing this area geographically are the supposed lower degree of agglomeration and a relatively one-sided, less advanced economic structure outside the West. This book, however, demonstrates (in Chapters 4, 6 and 9) that this argument is untenable.

The structure-bolstering projects apparently comprise the regional cooperative projects for supply and subcontracting, logistics management, training and technology projects. For De Randstad (Noord-Holland, Zuid-Holland and Utrecht) and the small, contiguous provinces of Zeeland and Flevoland, support for this type of project is not forthcoming. The infrastructural bottlenecks that appear there in particular are tackled by the Ministry of Transportation and Waterways, guided by a so-called mobility scenario. Some of the large industrial firms located there receive direct support from the Ministry of Economic Affairs (e.g. state participation in the Fokker aircraft industry).

The connection between regional socioeconomic policy and spatial policy was ruptured in 1983 and was not reinstated in the reformulation memorandum of 1988. This memorandum employs the zonal principle, dividing the Netherlands into deprived areas (peripheral or undergoing restructuring), intermediate areas (no high unemployment, support for structure-bolstering projects) and the De Randstad and its immediate vicinity (on the average, a favorable economic structure and hence ineligible for support). The Fourth Memorandum for Physical Planning (1988) adopts the nodal principle instead, which embodies the concept of urban clusters, each of which has high-quality production, centers of learning (universities) and top-ranking facilities (cultural, medical and retail). De Randstad, with four top centers, and an intermediate zone (extending to the urban clusters of Eindhoven and Arnhem-Nijmegen) is considered to be one single urban field. The differentiation between the West and the rest of the Netherlands is not found in the Fourth Memorandum, but it is found in the reformulation memorandum for Regional Socioeconomic Policy.

Outside the urban field, three urban clusters are distinguished in the Fourth Memorandum, namely Enschede-Hengelo (Twente), Maastricht-Heerlen (Zuid-Limburg) and Groningen (the North), which may serve as growth poles for the deprived areas. In these clusters, a better balance between spatial and regional socioeconomic policy should be feasible, but the reformulation memorandum neglects to acknowledge a nodal structure within the zonal regional-ization.

A creative debate on the relation between spatial and regional socioeconomic policy is imperative in the 1990s. It should treat other issues besides the first track of regional policy, namely the support for deprived areas. Moreover, the attention should be focused on the creation, also by way of physical planning activities, of good conditions for projects that contribute to macro-economic growth.

6 Economic Revitalization and the Region

6.1 Introduction

Chapter 3 elaborated on the significance of renewal in the economy of the Netherlands. In that context, it referred to the reports of the Commission for the Progress of Industrial Policy (the Wagner Commission). This organ listed a number of economic activities which had the best prospects for the near future. Since this was not part of the commission's mandate, however, the Wagner Commission did not deal with the spatial element.

The spatial aspect of the revitalization of the Dutch economy is the focal point of this chapter. Renewal is described on the basis of five different indicators: the spatial distribution of firms in industrial sectors with good prospects; the innovative nature of regional production environments; the spatial distribution of firms involved in certain kinds of innovation; the distribution of alert firms; and the regional variation in the number of newly established firms. These five indicators differ greatly. The majority of the starters, for example, are not in the least involved in high-tech industries or activities with good prospects.

6.2 Activities with Good Prospects

It is not easy to identify the location of firms specializing in activities with good prospects. The problem is that the activities thus defined (for industry, by the Wagner Commission; for services, by the Steering Committee Services Research, also called the Oostenbrink Commission) have to be translated in industry codes in order to make use of data from the Central Bureau of Statistics (CBS). The available CBS data provide much less detail than the classification of the Wagner Commission and the Oostenbrink Commission, however. In addition, the CBS categories are at the level of firms, not individual products. Moreover, the categories reflect the familiar past more than the unknown future. An

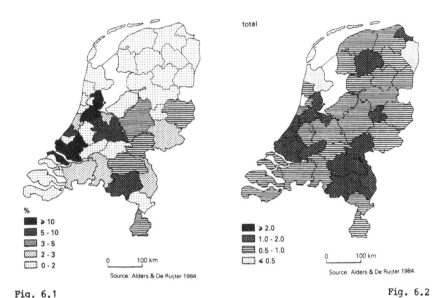

total

%
■ ≥ 10
▨ 5 - 10
▤ 3 - 5
▨ 2 - 3
▦ 0 - 2

0 ____ 100 km

Source: Alders & De Ruijter 1984.

■ ≥ 2.0
▨ 1.0 - 2.0
▤ 0.5 - 1.0
▦ ≤ 0.5

0 ____ 100 km

Source: Alders & De Ruijter 1984.

Fig. 6.1

Fig. 6.2

Employment (as a percentage of total) in sectors with good prospects (excluding agriculture), 1982

Employment in sectors with good prospects (location quotients), 1982

activity with good prospects such as 'electronic media' is indeed not easy to translate into one of the existing industry codes. Therefore, empirical research has no recourse but to use industry groups that reflect as well as possible the activities with good prospects. Many firms belonging to such an industry group in fact do not have good prospects at all. And the reverse is also possible; many firms with good prospects are listed among industry groups that are not considered promising.

When, for the sake of convenience, this 'translation' problem is ignored, then the TNO study (Alders & De Ruijter 1984) provides a clear picture of the spatial pattern of production activities in the Netherlands that have good prospects. The reports by the Wagner Commission and the Oostenbrink Commission permit us to identify 110 promising subgroups of industry and services (excluding agriculture). This produced the picture shown in Fig. 6.1. Five COROP regions have a high level of employment in firms with good prospects: Greater Amsterdam, Rijnmond, the agglomeration of The Hague, the province of Utrecht and the southeastern part of the province of Noord-Brabant (the region of Eindhoven). Nonetheless, a different picture emerges when employment is related to the size of the regions' population. Fig. 6.2

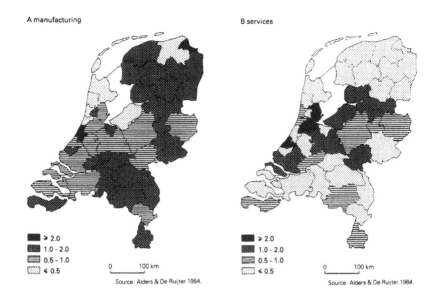

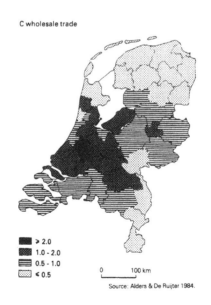

Fig. 6.3 Employment in sectors with good prospects in
manufacturing (a); services (b) and wholesale
activities (c), location quotients, 1982

shows remarkably little variation. The Netherlands apparently has no definite 'silicon valleys'. Manufacturing in the Netherlands is more widely dispersed than the service sector. It is still spread-out when differentiated according to high-potential industries, the wholesale branch and services (Fig. 6.3). Relatively much employment in high-potential industry is located in the periphery of the Netherlands (Fig. 6.3a). It is not surprising that the West has a strong position in the service sector, namely in the large agglomerations. In the more peripheral areas, the high-potential services are underrepresented (Fig. 6.3b). The West also scores relatively high in wholesale firms specializing in fields with good prospects, but the large agglomerations do not. The central part of the country is strong in this industrial sector (Fig. 6.3c).

From these patterns, it may be inferred that high-potential industry tends to locate where manufacturing is already well-established, and services with good prospects are found where the service sector is active. This pattern is clearer when, instead of using aggregated sectors (manufacturing, services, wholesale), industry groups (metals, printing) are used. Nearly every region has one or more sectors with good prospects. Friesland, for example, has much high-potential employment in food and kindred products, and Noord-Limburg has much metal industry with good prospects. In general, the areas in the West and the central part of the Netherlands have a less one-sided orientation toward (a single branch of) high-potential activities than the more peripheral areas. Yet in contrast to popular conception, there is no reason to assume that industry in the West is of relatively high value and high potential and forms a core area, whereas the peripheral areas specialize in low-caliber, unpromising sectors. Of course, the spatial pattern presented here is extremely dependent on the definition of high-potential categories of production. Therefore, a brief description of an investigation of the high-tech sector of manufacturing illustrates the classification of promising sectors. The researchers (Bouman et al. 1985) identified ten manufacturing subgroups. Since high-tech usually implies high-potential (the opposite is not true), nine of these ten subgroups were also included in the above-mentioned TNO investigation. The high-tech industry appears to be overrepresented in De Randstad. This is the location of 38.7 percent of all industrial and 42.3 percent of all high-tech firms. Even more noticeable is the fact that the periphery also has an overrepresentation of high-tech firms (Table 6.1). In contrast to Bouman et al., Koerhuis & Cnossen (1982) focused on one promising service sector, software and computer service firms. As expected, De Randstad shows an overrepresentation. The COROP region of eastern Zuid-Hol-

Table 6.1 Regional distribution of 'high tech' industries compared to manufacturing or industry in general (number of establishments).

	'high tech'		manufacturing (total)		industry (total)	
	abs.	%	abs.	%	abs.	%
Randstad	545	42.3	17,953	38.7	223,942	43.0
Intermediate zone	374	29.1	16,813	36.2	175,033	29.7
Periphery	368	28.6	11,641	25.1	121,659	27.3
Total	1287	100.0	46,407	100.0	520,634	100.0

Source: Bouman et al. (1985)

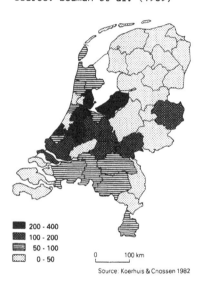

200 - 400
100 - 200
50 - 100
0 - 50

0   100 km

Source: Koerhuis & Cnossen 1982

Fig. 6.4 Over- and underrepresentation of software and computer service firms, 1981 (concentration ratios)

land, located in the Green Heart, has the highest concentration value, though (Fig. 6. 4). Comparison of the center, intermediate zone and periphery on the basis of these data reveals that the service sector with good prospects tends to be located in the West much more than the high-potential manufacturing does (cf. Fig. 6. 3).

6.3 Innovative Production Environments

Chapter 3 referred indirectly to the notion of 'production environment' in the context of the product life cycle. De Smidt (1981) describes it as: 'The whole of external

111

conditions outside a firm's own scope of influence that have proven their importance for the firm's location and financial operation'. Lambooy (1985) distinguishes three aspects of the production environment:
1   market relations, e.g. the position of a firm on the labor, the real estate and capital markets;
2   institutional relations with trade unions, political parties, environmentalist organizations, religious groups, etc.;
3   conditions: the presence of facilities (like infrastructure) without which firms cannot function.

Certain activities develop in some production environments better than in others. Oil refineries make different locational demands than furniture factories, for example, and high-tech firms require other conditions than apparel workshops. A production environment can have a stimulating or a debilitating impact on entrepreneurs' plans through each of the aspects named by Lambooy. The presence of high-potential activities in a given area may thus be related to the production environment there.

    An investigation of the economic potential of the regions in the Netherlands (NEI 1984) deals explicitly with the question of which production environments would stimulate innovative or high-potential activities. The innovation capacity of a production environment was tied to the presence of centers of learning (universities, polytechnics, laboratories), a highly qualified labor force, good

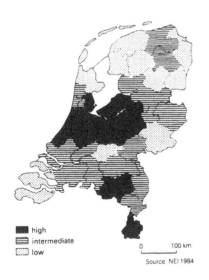

high
intermediate
low

0   100 km

Source: NEI 1984

Fig. 6.5   Innovation profile of the regional production environment

domestic and international access (airports), good and diversified industrial buildings, economies of agglomeration, an attractive residential environment, and a favorable investment climate. The choice of these indicators is related to the fact that revitalization is based on information and contacts. Partly for this reason, metropolitan production environments are considered to exert the most stimulating influence, which is also shown in Fig. 6.5, where De Randstad, but surprisingly also a traditional problem area like Zuid-Limburg, scores high.

The presence of centers of learning and large cities plays a central role in the presumed relation between production environment and economic renewal. It is implicitly accepted that the presence of a center of learning has a positive effect on the region. The same assumption is found in the geographical interpretation of the well-known growth pole theory (Vanneste 1967). According to this theory, technical, income and psychological polarization effects would especially benefit the region where the growth pole, the 'engine', is located (geographical polarization). Here we briefly treat the relation between production environment and good prospects. This chapter continues with a discussion of the significance of the production environment for industrial innovation and for the number of newly established firms.

Centers of learning are usually located in population concentrations. This is one of the reasons to assume that large cities provide a favorable production environment for new products and production processes. Empirical research, however, presents contradictory views on the spatial impact of centers of learning. Some investigators assign a decisive role to the presence of such centers. Others doubt that distance forms a barrier to the transfer of knowledge. These differences may possibly be related to the spatial scale under consideration. The notion 'metropolitan production environment' may be interpreted differently from one country to the next. In many areas in Western Europe, 'metropolitan zones' is a more appropriate concept than 'cities'. In a small, highly urbanized and relatively homogeneous nation like the Netherlands, the 'metropolitan zone' could in fact apply to most of the country.

In the Netherlands, investigations have been carried out on the regional significance of centers of learning for smaller industrial firms and the degree to which the factor of distance forms a barrier to the transfer of knowledge for these firms. Vlessert and Bartels (1985) investigated the extent to which differences in access to technological know-how existed between manufacturing firms located in regions with an unequal infrastructure of knowledge (the North, Zeeland and the area within a radius of 50 km around

Eindhoven). As they concluded, 'it appears that the more dynamic firms do not allow distance to influence their acquisition of information in any way, as observed in the Netherlands'.

Kok *et al.* (1985) arrived at a similar conclusion in an investigation of the renewal of products, processes, organizations and markets among small and medium-sized enterprises. Like Stokman (1986), they also found that the metropolitan production environments (the metropolitan areas of Rotterdam, Utrecht-Zeist and Arnhem-Nijmegen) were not characterized by relatively more innovative entrepreneurs. They concluded that 'the innovation profile of the metropolitan environment is therefore indeed more one-sided than the profile that emerges in the national picture' (Kok *et al.* 1985: 48). Furthermore, they propose that 'not only the presence of other firms but also other aspects which may be discerned in the metropolitan production environment (infrastructure, presence of facilities, socio-cultural amenities, contacts with the government) are not really considered to be important to the process of innovation. The only exception is the aspect of the labor market' (Kok *et al.* 1985: 99). In line with the finding of Vlessert and Bartels (1985), the researchers had the impression 'that the innovative enterprise obtained information wherever it was available and that distance played a very limited role in this regard' (Vlessert & Bartels 1985: 77).

These results relativate the distance barriers to obtaining new sources of knowledge in a small country like the Netherlands. On the other hand, the necessity to locate in the vicinity of a center of learning depends on how sensitive a given firm is to information. Although the above conclusions refer to existing smaller and not specifically information-sensitive firms, they may also apply to high-tech firms. This is indicated by the high-tech firms that have been established in the vicinity of a technical university by ex-staff members or ex-students, though factors other than proximity to the technical university may have played a role in the location choice. In general, starters tend to begin an enterprise in the region where they reside. This is also suggested by Bouman *et al.* (1985) in their investigation of high-tech industries. They observed a relation between high-tech and degree of urbanization. The rural municipalities and the urbanized rural municipalities showed a clear underrepresentation of high-tech firms. The overrepresentation in cities found in their investigation, however, does not correspond to the classification in center and periphery. Cities in provinces such as Gelderland, Noord-Brabant and Limburg also scored well.

In the studies described up to this point, the

114

Table 6.2   Demand profile of software and computer service firms in the Netherlands as of 1 June, 1981.

Locational factors	average	variance
1. Centrality in the Netherlands	3.27	0.58
2. Pleasant residential and living environment	3.21	0.55
3. Parking facilities	3.19	0.55
4. Price of land and buildings	3.17	0.83
5. Local traffic situation	3.16	0.60
6. Availability of trained personnel	3.04	0.74
7. Quality of available buildings	3.00	0.72
8. Housing options for personnel	2.98	0.70
9. Presence of a local demand	2.90	1.00
10. Option to expand	2.86	0.72
11. Size of available buildings	2.75	0.81
12. Possibility of subsidy	2.66	1.02
13. Government contracts and commissions	2.65	1.03
14. General familiarity with business community	2.56	1.03
15. Quality of the surroundings of available premises	2.53	0.92
16. Sales to parent company	2.34	2.86
17. Quality of available premises	2.31	1.03
18. Presence of supporting establishment	2.31	0.81
19. Access to scientific expertise at (technical) universities	2.10	0.90
20. Size of available premises	2.10	0.95
21. Presence of hardware suppliers	2.09	0.82
22. Acquaintance with center for micro-electronics	2.09	0.99
23. Acquaintance with related firms	1.93	0.75
24. Presence of related firms	1.81	0.73

Source: Koerhuis & Cnossen 1982, 120

importance of production environments for high-potential or innovative industrial activity is indirectly determined by reviewing the nature of the location of these firms. Yet the importance of the various locational factors can also be determined directly by interviewing the entrepreneurs. This has been done by Koerhuis and Cnossen (1982) for software and computer service firms. The results are here combined in the demand profile of this sector (Table 6.2). The important locational factors include some that also apply to enterprises in general (which is demonstrated in studies of firm relocation). Among these are parking facilities, price of land and buildings, local traffic situation, quality of the buildings, etc. 'Centrality in the Netherlands' had a notably high score. This factor scored even higher than 'presence of a local demand', despite the fact that the activities in this particular branch should be strongly oriented toward their customers, in view of their stage in the life cycle. This again is probably related to the short distances in the Netherlands. Great importance is attached to a 'pleasant residential and living environment', a result that is also found in studies from other countries. This is related to the shift of high-value activities toward the south in Great Britain, West Germany and France. A third important factor is the 'availability of trained personnel'. This is a locational factor for which regional differences continue to occur in the Netherlands.

Bouman *et al.* (1985) also queried the firms about their locational requirements. On the basis of the literature from other countries, they assumed that four locational factors would be decisive for firms in the high-tech sector: availability of highly qualified personnel, infrastructure of knowledge (e. g. presence of a (technical) university), the availability of venture capital, and the attractiveness of the residential/work environment. Two of these factors scored high in the investigation of Koerhuis and Cnossen (1982). Yet Bouman *et al.* (1985) found that the last factor, the attractiveness of the residential/work environment, was not very important. Greater significance was given to the presence of highly trained personnel and of centers of learning. This discrepancy may be due to the fact that Koerhuis and Cnossen investigated service activities and Bouman *et al.* were concerned with manufacturing. On the other hand, these results illustrate the lack of clarity that still exists with regard to the relation between production environment and high-tech activities.

6.4 Innovative Industry

The discussion of high-potential industry is based on the presence of establishments that have been allocated a 'high-potential' industry code. Because, as remarked earlier, not every enterprise in a high- potential branch has good prospects for the future, we now discuss innovation in business. The emphasis is on existing establishments; new enterprises will be dealt with later.

Interesting results are derived from an investigation commissioned by the National Physical Planning Agency and carried out by Kok *et al.* (1984). This investigation applies a broad definition of 'innovation'. Besides primary innovations (those which were previously unknown in the Netherlands and elsewhere), secondary and tertiary innovations were also included. Secondary innovations are items that are new to the Dutch economy but were already found elsewhere. Tertiary innovations are new to a given firm. Besides product innovation, following Schumpeter, process-, market- and organization-related innovations were distinguished.

By using several indicators, the investigators attempted to measure the propensity to innovate in existing firms. In the first place, they used data on firms that

Table 6.3   Innovation indices by province.

indicator:	InNu	Pil	patent	sample SME	
	location quotients by province				
job category:	0-100	0-500	all	0-100	
SIC category:	2,3,5	2 & 3	2 & 3	2&3,61/62,76,84	
province:				all innova- tions	only primary and secondary innovations
Groningen	76	94	73	113	79
Friesland	86	84	131	83	95
Drenthe	81	98	89	91	-
Overijssel	92	79	104	140	244
Gelderland	117	124	114	127	135
Utrecht	119	227	125	80	51
Noord-Holland	111	43	68	85	52
Zuid-Holland	99	101	108	106	139
Zeeland	86	134	73	50	68
Noord-Brabant	113	107	128	78	78
Limburg	81	106	62	126	60

Source: Kok et al. 1984, 69

participated in the Industrial Innovation Project (IIP), which has been discontinued in the meantime. The IIP had two target groups: starting manufacturing firms that wanted to exploit a new invention, and already existing firms that wanted to renew their product line. Kok et al. (1984) dealt with the latter category. Data were also derived from a representative questionnaire of firms with fewer than 100 employees in manufacturing, wholesale and supporting enterprises in transport and business services. A third indicator was formed by the patent requests recorded from August 1982 to August 1983. The last indicator was formed by the firms that had shown interest in the national program 'Innovation Now'.

The results are presented in Table 6.3 as concentration scores. The peripheral provinces demonstrate a low innovation propensity, which confirms the expectations based on the theory of regional-economic development. Yet the lag is actually not too great. In 20 percent of the cases, a peripheral province even has a concentration score of over 100. A high propensity to innovate is found for Noord-Brabant, Gelderland and Overijssel, which the researchers call the 'halfway zone'. They even pose that 'Gelderland has concentration scores of over 100 across the board and thereby shows itself to be our most innovative province.'

This leaves only the West, which should have the highest level of innovation on the grounds of agglomeration economies. This is not the case, however. Zuid-Holland scores positive, Utrecht shows fluctuating values and Noord-Holland is 'without a doubt the least innovative province of the Netherlands'.

From this conclusion, we could infer that the differences in innovation propensity in business are not large in the Netherlands. This may be due to the fact that access to information (which is essential for innovation) is relatively good throughout the Netherlands. The regional variation in the susceptibility of entrepreneurs to innovation may be low. Kleinknecht and Mouwen (1985) confirm that in a country like the Netherlands, a close relation should not be presumed between the way firms function and the locale of the regional production environment. They interviewed entrepreneurs at 1,842 manufacturing firms, of which most had more than 50 employees. They too assumed that in urbanized areas the conditions for innovation would be most favorable. All municipalities were then assigned an agglomeration score (Fig. 6.6). This procedure took the population size into account, as well as the distance to the other large municipalities. Thus a hierarchy was developed running from agglomeration index 1 (Amsterdam, The Hague, Rotterdam, Utrecht) to 9 (Leeuwarden, Emmen, Terneuzen).

The entrepreneurs were queried on their familiarity with a variety of government programs to promote innovation and on the use they had made of these. Entrepreneurs in the most urbanized areas prove certainly not better, and more often less, well-informed about the various programs than entrepreneurs in the so-called halfway zone. Some examples are as follows: of all firms in the biggest agglomerations 73.2 percent had heard about the existence of the technical development credit scheme and 14.4 percent had used this innovation incentive. In the most rural areas of the periphery (agglomeration index 9), these figures were 83.0 percent and 18.0 percent. The least urbanized areas also achieved the highest scores for the innovative management incentives. By most measures, the areas with an agglomeration index of 4, 5 or 6 scored best. This result is consistent with the findings of Kok *et al.* (1985).

An interesting addition is found in Stokman (1986), who differentiates a traditional (SIC industry group 20-27) and a modern sector (SIC 28-39) among small and medium-sized enterprises (SME). The intensity of research and development (R&D man-years in 1983 as percentage of total employment at year-end 1983) is, as expected, highest in the modern sector (Table 6.4). It is interesting to note, however, that in the traditional sector, the firms in the periphery have the greatest R&D intensity. In the modern sector, the halfway

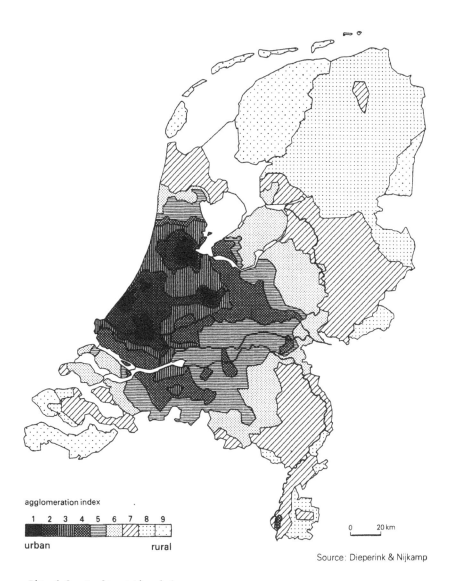

agglomeration index

| 1 | 2 | 3 | 4 | 5 | 6 | 7 | 8 | 9 |

urban   rural

0 ___ 20 km

Source: Dieperink & Nijkamp

Fig. 6.6   Agglomeration index

zone scores highest. In both cases, the firms in De Randstad
have the lowest R&D intensity. In terms of the nature of the
innovations (Table 6.5), the picture improves for De
Randstad, but the halfway zone remains in the lead. The
Randstad takes the lead only in truly new innovations (Table
6.6). But the firms in the periphery show remarkably good
scores on that point as well.

Table 6.4   Regional distribution of traditional and modern small and medium-sized manufacturing enterprises according to R&D intensity rate* (in percentages).

	TRADITIONAL MANUFACTURING				
	none	0-2%	2-5%	5-8%	8% or more
Randstad	54.0	23.0	15.1	5.6	2.4
Intermediate zone	44.3	25.0	22.4	4.8	3.5
Periphery	48.0	22.0	19.0	5.0	6.0
Total	47.8	23.8	19.6	5.1	3.7

	MODERN MANUFACTURING				
	none	0-2%	2-5%	5-8%	8% or more
Randstad	43.9	21.5	19.7	7.2	7.6
Halfwegzone zone	35.0	18.8	21.7	12.3	12.3
Periphery	33.3	24.7	24.2	9.6	8.1
Total	37.3	21.2	21.8	10.0	9.7

* R&D intensity is defined as number of R&D man-years in 1983 as a percentage of total jobs (beginning 1984) in a firm.

Source: Stokman 1986

Table 6.5   Regional distribution of the average number of innovations realized, by firm, for traditional and modern small and medium-sized manufacturing enterprises.

	TRADITIONAL MANUFACTURING			
	product	process	combined	total
Randstad	0.94	0.41	0.38	1.74
Intermediate zone	1.12	0.44	0.28	1.84
Periphery	0.44	0.37	0.19	1.00
Total	0.92	0.42	0.29	1.63

	MODERN MANUFACTURING			
	product	process	combined	total
Randstad	1.07	0.42	0.20	1.69
Intermediate zone	1.29	0.49	0.29	2.07
Periphery	0.89	0.52	0.19	1.60
Total	1.12	0.48	0.23	1.83

Source: Stokman 1986

120

Table 6.6   Innovations new to the industry. Share of total innovations
            in small and medium-sized manufacturing industries, by
            region and type. (in percentages)

	product	process	combined	total
Randstad	73.2	40.7	41.5	60.7
Intermediate zone	52.3	33.7	26.1	44.1
Periphery	70.1	36.7	47.5	55.1
Total	61.3	36.4	34.4	50.9

Source: Stokman 1986

Davelaar and Nijkamp (1987) reach a similar conclusion, namely that the regional differences in innovation propensity are relatively minor. They distinguish product innovation and process innovation. According to them, the central zone (De Randstad) appears to launch relatively more product innovations of a creative nature (i. e. new to a given branch of industry). For creative process innovations, the intermediate or halfway zone scores highest. 'It appears that a greater overall propensity to introduce process innovation exists in the periphery' (Davelaar & Nijkamp 1987: 722).

6.5 The Alert Enterprise

The fourth indicator of the dynamics of regional economies employed here is the location of (recent) award-winning firms and the place where subsidies for improvement of the quality of enterprises are allocated.

The national 'awards' used in this context include export awards, awards for good financial reports, for innovative entrepreneurship, etc. Further, we included the 22 firms that were admitted, after a rigorous selection procedure, to the 1986 research program for small and medium-sized enterprises of the Association for Technical Sciences (Stichting voor de Technische Wetenschappen STW). Finally, we checked the location of 25 smaller enterprises that had shown spectacular growth in terms of sales (the 'biggest little growers'), as published in the periodical Inter-magazine. Fig. 6.7 demonstrates that award winners are found throughout the Netherlands. Of course, this indicator is not above criticism, but it seems at least plausible that the firms it embodies are not the worst. Keeping this in mind, an interesting picture emerges when we relate these 'winners' to the total regional population of firms. Then De Randstad provinces of Noord-Holland and Zuid-Holland prove to lag behind Noord-Brabant, located in the halfway zone,

Source. Horvers & Wever 1987

Fig. 6.7   Locational pattern of award-winning firms

and even more behind the peripheral provinces of Groningen and Drenthe.

Of the many possible subsidy programs, three are discussed here: the technical development credit scheme (TOK) the subsidy program for management support (SMO) and the innovation incentive program (INSTIR). The technical development credit is a risk-bearing loan for enterprises that intend to develop new ideas or discoveries. In order to become eligible for such a loan, the project submitted for review must be, entirely or for the most part, technically new in the Netherlands and must be significant to the Dutch economy (Horvers & Wever 1987). The subsidy program for management support permits a financial compensation for costs incurred by consulting a third party in regard to business practices, management counseling (for young firms) or automation. And lastly, the innovation incentive program subsidizes the wage expense of (technological) research. Leaving any possible disadvantage of this indicator aside for the moment, it is clear that firms that take the initiative, of their own volition or upon the advice of informa-

Table 6.7 Provincial location quotients by subsidy program: TOK, SMO, INSTIR.

Province	TOK	SMO	INSTIR
Groningen	1.09	1.09	1.03
Friesland	1.15	1.53	1.13
Drenthe	1.27	1.38	0.78
Overijssel	0.88	1.22	1.41
Gelderland	1.16	1.28	1.23
Utrecht	1.23	1.00	1.19
Noord-Holland	0.69	0.97	0.77
Zuid-Holland	0.96	0.71	0.90
Zeeland	0.48	0.65	0.76
Noord-Brabant	1.03	1.08	0.99
Limburg	1.45	1.03	0.86

Note:
TOK : This program is mostly used by manufacturing firms. Therefore, the location quotient is based on the provinces' share in the total number of manufacturing firms. The data refer to the period 1981 through 1985.
SMO : period Oct. 1984-Oct. 1985
INSTIR: period Oct. 1984-Oct. 1985

tion agencies, to participate in one of these programs at least have a view to the future. In that sense, the behavior of these firms is in line with ideas about regional potentials.

Table 6.7 gives the location quotients by province in which the share in the number of allocations to the three subsidy programs is related to the share in the total number of establishments as of 1 January 1985. Review of the data reaffirms the unspectacular findings for Noord-Holland and Zuid-Holland. Besides Gelderland and Utrecht, the provinces of Groningen and Friesland also show high scores for all of the programs. Drenthe does not do badly either, especially in comparison with the other 'small' province, Zeeland. Nowhere in the Netherlands is the production environment an obstacle to the establishment of good firms. This topic will be dealt with in the next chapter.

6.6 New Enterprises

The previous section demonstrates that the number of activities with good prospects for the future differs rather widely by province and that manufacturing firms are more involved in innovation in some areas than in others. A difference in regional dynamics is also observed in the number of newly started firms.

New firms are 'in' once again, after a period when the large corporation was the focus of attention. Many, in fact, have even put all their hope in new firms and starters to create jobs, because the large corporation tends to shed labor through automation. Many see an important role for the new and/or the small enterprise in developing new products and production processes, since such enterprises are more flexible than the large corporation and would thus be able to respond to new trends in the market.

Many factors play a role when someone decides to start a company of his or her own. In the first place, there must be an impetus, which Shapero (1983) called the displacement factor. In a country like the Netherlands, where most people are employed for wages, it is not common for an individual to trade the security offered by the Welfare State for the risks of enterprise. The factors that influence people to take that risk include dissatisfaction with the job they have, a sense of being undervalued and impending or actual unemployment. Yet not everyone who is unemployed or dissatisfied will start their own business. This leads us to the second factor, the disposition to act, which refers to the character of potential starters. People who take the step to entrepreneurship typically desire to be independent. But then again, not everyone with this type of character will take impending unemployment as the cue to start a business of their own. The social environment also plays a role, constituting the 'credibility factor'. Entrepreneurship is not held in high esteem everywhere or by everyone. A remark by Shapero puts this succinctly: 'In Italy I found that a man of education who started a business lost social status, in the US that man is a folk-hero'. Without a doubt, many firms are established by people whose families are already in business and thus enjoy a high degree of credibility in their efforts. They are acquainted with entrepreneurship and find approval for their decision to start their own business from the people in their social circles.

Shapero calls the fourth factor the 'availability of resources'. This includes the impetus offered by the production environment: the presence of suitable and inexpensive housing, taxation rates, information opportunities and quality, size of the market, financing options, etc.

There is no guarantee that an entrepreneur starting out in business will be successful, no matter how highly motivated he or she is. In the Netherlands, rougly 16,000 businesses are set up each year. After five years, more than 50 percent of these are defunct (Fig. 6.8). After 15 years, at most 25 percent are still in operation. A multitude of factors determine the degree of success. One factor is the product-market combination. By far not everyone is able to find the elusive gap in the market. With some products, even the best businessman will only be able to break even. In

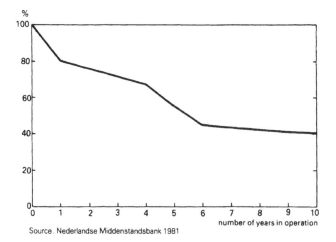

Source. Nederlandse Middenstandsbank 1981

Fig. 6.8   Survival of newly established firms in the
            first ten years of their existence
            (percentages)

branches with an overcapacity, one starter's 'drink' is
another, already established, entrepreneur's 'poison'.
Still, new firms making that product keep appearing on the
scene, especially when the threshold (comprising requisite
diplomas, permits and the size of the investment) for entry
is low, because every starter believes that he or she will
succeed where the others have failed. Reality is different,
however. There are countless cases of disappointment after
an enthusiastic opening of a bar, bistro, sport school,
video rental or sauna. Besides the product-market combination, other factors play a role. These include the initial
financial basis of the enterprise, the extent of prepara-
tion, the entrepreneur's knowledge of the field, his or her
persistence, the support that can be mustered within the
family, etc.

Given the factors that influence starting up in
business, the number of newly established enterprises may be
expected to differ by region. The impact of the 'displacement factor' may be different. At any rate, unemployment, or
the threat of it, is higher in some areas than in others. It
is even conceivable that dissatisfaction with a current job
is greater in regions where production work is monotonous.
The 'disposition to act' may also differ by region. It is
possible that migration plays a role. Migration is selective
by nature and this could contribute to the loss of a

125

region's most energetic individuals, possibly those who are most liable to start a business.

The 'credibility factor' is partially formed by the structure of employment, which differs regionally. In areas where employment is concentrated in a few large firms, there are fewer 'role models' for a starter than in an area hosting many small companies. In areas with much 'production work', the point of departure seems to be less favorable, since experience in monotonous work is not the ideal background for independent entrepreneurship. In that connection, working in a small enterprise, where one is confronted with many aspects of business, is more advantageous. In addition, sectors have different entrance thresholds. Someone with a great deal of experience in a capital-intensive industry (e.g. oil refinery) is not likely to start a new firm in that sector. On the one hand, huge investments would be required, and on the other hand, it is not easy to penetrate that market. Individuals from such sectors are virtually forced to enter sectors with which they are less well acquainted, if they wish to start a business. Sectors such as wholesale, retail and personal services (pedicure, typing, translation) have a low entrance threshold. Of course, this is part of the reason why the number of new business service (including personal services) establishments is nearly twice the amount of new firms in manufacturing or industry, expressed as a percentage of the total number of existing firms.

There are also regional differences in 'availability of resources'. The classical incubation theory, developed in the US, provides some insight into these differences. According to Struyk and James (1976), small new enterprises are likely to locate in areas where the facilities they need to function are already in place (external economies). These facilities include suitable, inexpensive business premises, the presence of numerous other small firms (clients, suppliers, services), the presence of easily accessible information about the market and production costs, etc. Leone and Struyk (1976) summarize this theory in two hypotheses. The simple hypothesis poses that central locations will generate more new firms than other parts of a city. The complex hypothesis poses that new firms are created in central parts of a city, but that they leave this location after a few years in order to expand elsewhere. The central parts of the city serve as an 'incubator'.

Although the empirical testing of these hypotheses yielded few convincing findings (Jansen 1981, De Ruijter 1978), the incubation theory retains its popularity. In regional terms, this theory implies that the number of new firms decreases as the distance to a large city increases. On the basis of the presumed importance of external econo-

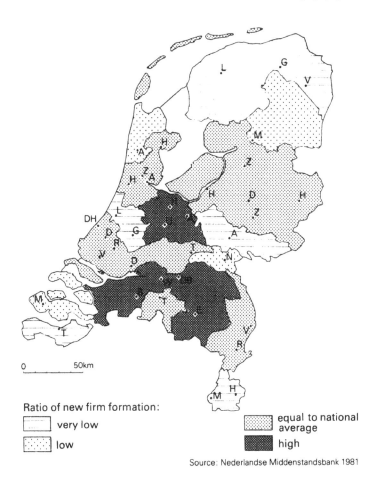

Ratio of new firm formation:

□ very low

▦ low

▦ equal to national average

■ high

Source: Nederlandse Middenstandsbank 1981

Fig. 6.9   The dynamics of new firm formation: firms
registered after 1970 and still in existence at
year-end 1979

mies, the purely rural areas would not be very 'fertile'
ground for new firms. This hypothesis, incidentally not yet
confirmed, has been formulated by Ten Heuvelhof and Musterd
(1983), among others.

There is little known about the number of newly
established firms per region in the Netherlands. This is in
part due to the difficulty of defining what is a new firm.
When is a new firm operational? The moment the first order
is received, the moment it is officially registered with the
Chamber of Commerce, or the moment the owner derives most of
his or her income from the business? It is common practice
to use the date of registration with the Chamber of
Commerce. On the basis of Chamber of Commerce registers, the

127

1980 financial report of the Nederlandse Middenstandsbank composed a picture of the distribution of new firms (Fig. 6. 9). This overview reflects the firms newly registered after 1970 which were still in operation at the end of 1979 in the sectors of manufacturing (SIC 2-3), trade (SIC 6, excl. 67) and business services (SIC 83. 2, 84 and 85). The similarity to the distribution of activities with good prospects and of innovation in existing businesses is remarkable. Here, too, the regional differences are minor. The North and Zuid- Limburg, the two traditional problem areas, score surprisingly low, however, and Noord-Brabant and the central part of the Netherlands again score higher than the cities in Noord-Holland and Zuid-Holland.

Table 6.8 Total number of starts* per 10,000 inhabitants and the share of starts in all manufacturing firms (%).

District	Total			Industry		
	1970	1975	1980	1970	1975	1980
Veendam	10.4	11.6	11.2	7.2	7.0	8.1
Heerlen	13.1	14.2	17.9	7.6	6.7	8.0
Amersfoort	17.5	26.7	31.0	7.7	8.9	7.7
Amsterdam	22.8	23.8	30.0	9.0	8.0	7.3
Rotterdam	14.5	23.4	28.5	8.5	7.7	9.2

* With the exception of hobby firms, statutory registrations and firms with no employees   Source: Wever 1984

The data in the financial report of the Nederlandse Middenstandsbank refer to the net number of establishments and closures. A more refined picture, including only starts, can be derived from the cohort analyses performed by Wever (1984, 1986), although these do not cover the entire country. A cohort analysis checks what has happened to the firms established in a given year. Such an analysis was performed for 25 of the 36 districts of the Chambers of Commerce in the Netherlands. For the sake of convenience, we only present the results for a limited number of areas here. These include a development area in the North (Chamber district Veendam), a restructuring area in the South (district Heerlen), an area in the halfway zone (district Amersfoort) and the two most definitely urban areas (Amsterdam and Rotterdam). The number of starts in these Chamber districts is shown in Table 6. 8. The picture thus sketched does not deviate much from the results presented by the Nederlandse Middenstandsbank in Fig. 6. 9, with the exception of the larger number of new establishments in Amsterdam and Rotterdam. The table shows a significantly large difference in regional dynamics per 10, 000 inhabitants, a difference that showed no tendency to decrease in the period 1970-1980. The number of starts in manufacturing as a percentage of the total varies considerably less. None of the five districts

Table 6.9 Defunct firms per cohort as percentage of starts.

District	1970	1975	1980
Veendam	61.7	52.5	41.9
Heerlen	60.9	52.6	21.4
Amersfoort	64.3	49.1	27.1
Amsterdam	71.2	59.0	32.1
Rotterdam	71.0	60.6	39.5

Source: Wever 1984

Table 6.10 New high-potential firms established in 1975 and 1980, by
region (total).

Region	1975			1980		
	A	B	C	A	B	C
North	8.4	0.92	0.86	8.4	1.29	0.74
South	7.1	1.11	0.71	9.1	1.67	0.80
Intermediate zone	9.5	2.15	0.95	13.0	3.65	1.14
Rijnmond	12.9	2.82	1.28	13.1	3.52	1.15
Randstad-North	9.7	2.35	0.97	10.3	2.84	0.91
The Netherlands	10.0	1.90		11.4	2.60	

A = new firms in high-potential sectors as percentage of all new
firms.
B = new firms in high-potential sectors per 10,000 inhabitants.
C = regional percentage of new firms in high-potential sectors as
related to the national percentage.

Source: Wever 1987, 172.

Table 6.11 New manufacturing firms in high-potential sector·
established in 1975 and 1980, by region: manufacturing.

Region	1975			1980		
	A	B	C	A	B	C
North	1.99	0.21	0.86	1.74	0.27	0.82
South	1.73	0.27	0.83	2.01	0.37	1.03
Intermediate zone	2.89	0.65	0.95	2.83	0.79	1.20
Rijnmond	2.54	0.55	1.36	2.92	0.79	1.39
Randstad-North	1.51	0.36	0.85	1.48	0.41	0.91
The Netherlands	2.11	0.40		2.07	0 47	

A = new firms in high-potential sectors as percentage of all new
firms.
B = new firms in high-potential sectors per 10,000 inhabitants.
C = regional percentage of new firms in high-potential sectors related
to the national percentage.

Source: Wever 1987, 175

129

Table 6.12 New firms established in 1980 in expanding and slightly
stagnating sectors (A) and in receding sectors (B), as
percentage of all new firms and of all new manufacturing
firms.

Region	of total		of manufacturing	
	A	B	A	B
North	3.6	1.8	38.2	19.1
South	3.4	2.0	39.5	22.9
Intermediate zone	3.9	2.3	38.6	22.6
Rijnmond	4.3	0.9	47.0	9.1
Randstad-North	3.0	1.8	42.1	25.4
The Netherlands	3.3	1.8	36.9	19.2

Source: Wever 1987, 176

registers even 10 percent. More recent (1985) data, though
not comparable with the figures from the Association of
Chambers of Commerce presented in Table 6.8, give the
impression that not much changed in the distribution after
1980. In 1985 as well, the total number of starts was
highest in the districts of Amersfoort and Rotterdam and
lowest in Veendam (no data were available for Amsterdam).

The observed deviant position of Rotterdam and Amsterdam in comparison with the findings presented by the Nederlandse Middenstandsbank is clearly expressed in Table 6.9.
Although many firms were established in the two most
urbanized districts, their chance of survival is relatively
small, causing a modest net result. Incidentally, in terms
of policy, such a position is far preferable to the position
of the peripheral district of Veendam, where there is an
absolute shortage of starts. It is easier to reduce the
demise of established firms than to raise the number of
starts.

As far as the revitalization of the regional economy is
concerned, not every new firm is equally important.
Therefore we determined the number of new establishments in
high-potential activities, as defined by the Wagner
Commission and the Oostenbrink Commission. For convenience,
the focus was on three areas: the development areas in the
North (Chamber districts of Leeuwarden, Veendam and Meppel),
the restructuring areas in the South (districts of Heerlen
and Maastricht), the halfway zone (intermediate districts
such as Breda, Waalwijk and Tiel) and parts of De Randstad
(districts Amsterdam/Haarlem and Rotterdam/Vlaardingen). The
regional variation is shown in Table 6.10. Per 10,000
inhabitants, the largest number of new high-potential firms
are established in Rijnmond, the northern wing of De
Randstad and the southern halfway zone. The problem areas
still lag behind. Yet when the share of new high- potential
firms in the total number of new firms is considered, the

differences are smaller. Obviously, rather than the wrong kind of firms, too few firms are started in the problem regions. The location quotient presents a similar picture. The halfway zone again scores surprisingly high for new high-potential firms.

The data in Table 6.11 refer exclusively to industry with good prospects. The halfway zone and Rijnmond score high in this context as well. The region comprising the northern wing of De Randstad lags noticeably behind. This is related to the weaker orientation of this area toward manufacturing. The region Randstad-North has a better score for establishments in high-potential service sectors.

An investigation carried out by the Economisch Instituut voor het Midden- en Kleinbedrijf (Webbink 1985b) provides an interesting addition to this picture. Reasoning in terms of the life cycle, all the subgroups of industry were classified as expanding, slightly stagnating, stagnating, saturated, slightly declining and declining sectors. This classification was based on data for the period 1978-1981 in regard to growth of domestic sales, growth of export, share of export in the total sales, growth in employment, growth of the number of firms, growth of value added, and growth of investment in fixed assets. For the period 1981-1983, the classification was tested for stability, and it proved to be reasonably satisfactory.

Table 6.12 gives the data for the two growth classes (expanding and slightly stagnating) and the two declining classes (slightly declining and declining). Once again, the proportion of the total number of new establishments formed by the growth classes is not much different from that of the declining classes. This is particularly clear when the region Rijnmond, which has by far the best scores, is not included in the picture. For the problem regions as well, it is clear that the problems are not due to the establishment of the wrong kind of firms.

6.7 Conclusion

Reviewing all the findings, it may be concluded that the regional differences in the distribution of manufacturing within the Netherlands are relatively minor. In contrast to the popular image, the greatest part of the Netherlands may be considered as one single urban field. The differences therein seem to have more to do with regional variations in the entrepreneurs' receptivity to renewal than with regional variations in the accessibility of information. At this point, the halfway zone seems to have the best prospects.

7 Corporate Internationalization: The Position of the Netherlands

7.1 International Position

The industrialization of the Netherlands is connected to the international business world in many ways. Some Dutch industrial corporations appear high on the world list of large enterprises. From the start, firms such as Shell and Unilever had a global field of operations. From the 1960s on, more and more Dutch manufacturing firms have gone multinational. They extended their reach to cover other continents; the United States was the most favored direction for growth. The background of this process of multinationalization will be discussed below (Section 7.2). Particular attention will be given to the rank of Dutch enterprises on the list of global corporations and of big European enterprises. The swelling stream of investment leaving the Netherlands and the flow of foreign investment into the country will also be discussed. Other characteristics of the big Dutch industrial corporations are treated in greater depth in Chapter 8.

Thus, the impact of international influences on the Dutch economy is the central theme of this chapter. Section 7.3 treats the importance of foreign enterprise for the Netherlands as a whole and, in particular, for the regions within the country. An interesting question in this context is how this international influence has changed in the course of time and what role the various regions have played in this process. Since the contribution of the region to national prosperity is a topic of major concern, it would be interesting to see if any functional-spatial specializations can be identified.

The Netherlands is considered to be a 'Gateway to Europe'. Yet within the Netherlands, there are two foci of intensive contacts with other countries: a seaport and an airport of international stature, namely Rotterdam and Schiphol. In Section 7.4, these two international transportation hubs are subjected to further examination from the perspective of industrialization.

The international influence on the Netherlands is thus

described by placing the Dutch corporations in their domestic sphere of influence, sketching the importance of foreign enterprise for the Dutch (regional) economy, and analysing the two foci of foreign influence within the 'Gateway to Europe'. This profile of the international influences on industrialization in the Netherlands may seem too narrow in scope. But it does complement the description of the competitive position of Dutch enterprise in the international arena, which was dealt with in Chapter 3. In addition, the top-ranking enterprises in Dutch industry will be treated in Chapter 8.

7.2 European Big Business:
Internationalization Tendencies and Structural Differences

In the mid-1960s the Editor in Chief of the prestigious French weekly L'Express, Jean Jacques Servan-Schreiber (1967) sounded the alarm in his book Le défi américain (The American Challenge). In his opinion, US corporate firms were definitely taking the lead in most advanced sectors of the economy, showing superiority in technology as well as in market strategies, management and even mentality. European firms had been pushed to the brink of total collapse. This pessimistic view could not take into account the hardships of economic stagnation and energy crises in the years that followed. How did large European corporations manage in the years between 1973, on the eve of the first oil crisis which marked the end of the long-lasting economic boom, and 1984, the year of the beginning of the structural recovery of the Western economies? Do the main countries of origin show any distinct patterns of multinational business activity in terms of size, sectors and locations of top management centers? Which changes in the geographical pattern of international direct investment were mainly controlled by multinational business? These are the central questions that will be answered in this section, taking into account the shortcomings of the available statistical documentation.

A Profile of Leading Corporations

The world stage of big business is dominated by US firms. On the basis of a volume of sales of at least 25 billion dollars in 1984, a list of fifteen top-ranking corporations can be drawn up from the *Fortune* list (Table 7.1). It is found that only three non-American firms (Shell, BP, ENI) qualify for this list, all three being oil companies. A cursory comparison with the ranking for 1973 gives the impression that the European position has indeed been worsening: e. g. Unilever dropped from ninth place in 1973 to eighteenth on the 1984 list, Philips from thirteenth to twenty-seventh. Suc-

Table 7.1 The world's biggest firms, 1973 and 1984*.

1973	1984
1. General Motors (US)	1. Exxon (US)
2. Exxon (US)	2. Shell (Netherlands, UK)
3. Ford (US)	3. General Motors (US)
4. Shell (Netherlands, UK)	4. Mobil (US)
5. Chrysler (US)	5. Ford (US)
6. GEC (UK)	6. British Petroleum (UK)
7. Texaco (US)	7. Texaco (US)
8. Mobil (US)	8. IBM (US)
9. Unilever (UK, Netherlands)	9. Dupont de Nemours (US)
10. IBM (US)	10. ATT (US)
11. ITT (US)	11. General Electric (US)
12. Gulf Oil (US)	12. Standard Oil of Indiana (US)
13. Philips (Netherlands)	13. Chevron (US)
14. British Petroleum (UK)	14. ENI (Italy)
15. Nippon Steel (Japan)	15. Atlantic Richfield (US)

* Rank according to sales (above 25 billion dollars in 1984, compared to the top 15 of 1973)

Source: Fortune 1974, 1985

Table 7.2 The biggest private enterprises in Europe, 1972-1984, by worldwide employment (x 1000)*.

	1984	1977	1972
1. Philips (Netherlands)	344	384	371
2. Unilever (UK, Netherlands)	319	327	337
3. Siemens (FRG)	319	319	301
4. Volkswagen (FRG)	238	192	192
5. Fiat (Italy)	231	139	190
6. British-American Tobacco (UK)	213	152	.
7. Daimler-Benz (FRG)	200	138	150
8. Peugot/Citroën (France)	188	185	.
9. Hoechst (FRG)	178	181	146
10. Bayer (FRG)	175	170	104
11. Shell (Netherlands, UK)	149	155	174

* By rank in 1984, with rank in 1977 and 1972 for comparison, for firms with more than ± 150,000 employees worldwide.

. Comparative data unavailable

Source: Fortune 1973, 1978, 1985

cessive energy crises caused oil and gas prices to escalate to such an extreme level that the volume of sales of oil companies grew much faster than could be achieved by companies in other branches. For that reason 'new' names of firms showed up in the 1984 list, mainly oil and chemical corporations (Dupont, Standard Oil of Indiana, Chevron, ENI, Atlantic Richfield).

A separate class of giant firms, with a sales volume of between 85 and 90 billion dollars in 1984, includes Exxon, Shell and General Motors (Ford fell into this category till the mid-1970s). The joint volume of their sales is comparable to twice the gross domestic product of the Netherlands. A second league of firms, each having a volume of sales of approximately 50 billion dollars, consists of three oil companies (Mobil, BP, Texaco), apart from Ford and IBM.

Ranking large corporations according to volume of sales gives a narrow view of the significance of firms. Employment numbers or assets are also important yardsticks. By comparison with car manufacturing or electronics firms, the volume of sales of oil companies exerts little influence on numbers of jobs. Though its volume of sales is five times as large as for Philips, Shell has 149,000 employees whereas Philips has 344,000.

Looking more specifically at the course of private European corporations during the decade of economic stagnation and energy crises, it turns out that German firms took a leading role in the European business arena. European firms which, at least on one occasion during this period, had 150,000 employees have been compared in Table 7.2. Leading German corporations in fields such as chemicals and pharmaceutical products (Hoechst, Bayer) maintained their employment growth of the mid-1970s. German car manufacturing has recovered recently (VW, Daimler-Benz). European top-ranking corporations, judging by employment figures, are strongly involved in car manufacturing, as is illustrated by Fiat (Italy) and Peugeot/Citroën (France).

Apart from German firms, the position of Anglo-Dutch corporations like Unilever and Shell is remarkable, as is Philips in terms of employment (cf. Siemens). The significance of British, French and, in particular, Italian firms might be underestimated since government-owned (or controlled) corporations have not been listed in Table 7.2. Comparisons for these firms could not be made for the long term.

British corporations occupy a leading position on the scene of European-based large corporations, ranked on *Fortune's* top-500 list of large non-US corporations. One out of every three jobs controlled by European firms is paid for by these British corporations. They are at the same time responsible for one-third of the total volume of sales (Table 7.3), measured on a global basis. German-based firms rank second, while French and Italian firms follow far behind.

Most remarkable is the ranking of some smaller nations, such as the Netherlands, Sweden and Switzerland. The relation, however, between volume of sales and number of jobs controlled is not the same for these countries. The Dutch firms show a high volume of sales compared to the number of

Table 7.3 The importance of big European firms, by location of head
office in 1984*.

Country	Number of firms	Sales worldwide (millions of dollars)	Number of jobs worldwide (x1000)
United Kingdom	77	323,111	3,770
Federal Republic of Germany	55	239,266	2,898
France	34	149,735	1,623***
Italy	12	83,588	1,070
The Netherlands	11	98,584	735
Belgium	5	19,611	93
Luxembourg	1	982	14
Denmark	0	0	0
Ireland	1	887	12
Greece	1	1,115	1
Spain	9	16,387	81
Portugal	1	2,161	7
Subtotal EC	205**	935,427	10,304
Sweden	17	41,473	564
Finland	7	12,219	116
Norway	3	9,682	35
Switzerland	12	50,576	602
Austria	4	9,361	87
Total Europe	248	1,058,738	11,708

*   Among the top 500 biggest non-American firms.
**   Shell and Unilever allocated to the Netherlands (Shell 60%,
      Unilever 40%) and to the United Kingdom (the remainder).
*** Correction for Saint Gobain (erroneously cited by Fortune).

Source: Fortune 1985

jobs. In the next section, 'national' structural differences
will be dealt with.

Both the UK and the Netherlands can be characterized as
the most highly multinationalized European countries (with
quotients of respectively 68 and 72 percent). France and
Italy show a lesser degree of multinationalization, while
Spain has the lowest, at 9 percent. Sweden and Switzerland
hold a stronger position than Germany in this respect.

Structural Differences between Large Corporations according to Country of Origin

A comparison of the structure of large corporations according to country of origin brings out some very interesting
contrasts. The most important European nations in this
respect can be compared with the aid of Fig. 7.1 and 7.2,
taking volume of sales and employment respectively, as
yardsticks.

The main contrast lies in two specific profiles of big
business: the UK versus the German 'model'. Of sales of

SALES

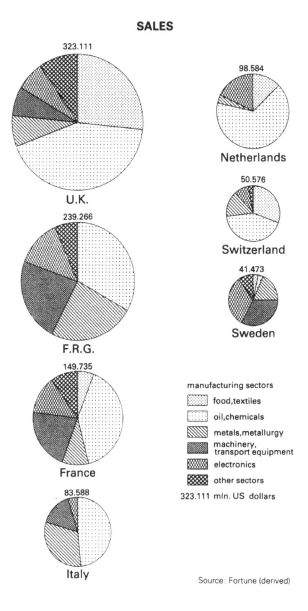

Fig. 7.1   Sales volume for companies of European origin
ranking among the top 500 manufacturing firms,
differentiated by sector, 1984

137

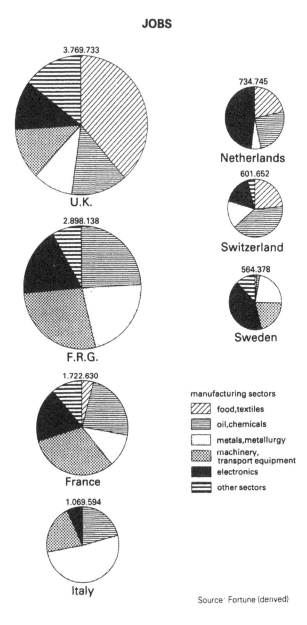

JOBS

Fig. 7.2 Employment in companies of European origin,
ranking among the top 500 manufacturing firms,
differentiated by sector, 1984

UK-based firms, 25 percent can be labeled as part of traditional sectors such as food and kindred products, textiles, etc. In Germany none of the fifty-five big corporations fall into these sectors! With respect to jobs, the impact is even stronger: nearly 40 percent of the employment in top UK firms is in these traditional sectors. By contrast, German firms have a strong foothold in two sectors, namely metals (including metallurgy) and machinery (including car manufacturing), controlling almost half of the total volume of both sales and employment in these sectors. Oil refineries and chemical industries occupy an important position in all industrialized countries. These large-scale investments are controlled by big private or state-owned enterprises. If measured by the volume of sales in these sectors, most industrial countries rate between 30 (Germany) and nearly 50 percent (Italy), with intermediate positions for the UK, France and Switzerland (pharmaceutical industry).

Exceptions at both ends of the scale are the Netherlands (65 percent) and Sweden (3 percent). This can be explained, as far as the Netherlands is concerned, by the impact of the giant Shell corporation (Shell accrues for 60 percent to the Netherlands and 40 percent to the UK). In terms of employment, this sector is of far less importance, mostly rating between 15 and 25 percent. Switzerland has highly specialized pharmaceutical industries which account for 40 percent of the total number of jobs in large corporations. Sweden, also highly qualified, is specialized in electronics and machinery.

The UK 'model', with its concentration on traditional consumer sectors, and the German 'model', characterized by production of capital goods and durable consumer goods, can be analysed in greater depth by looking at the ratio of sales to employment. The German-based chemical industries are not as specialized as the Swiss, while they are not as strongly oriented toward bulk chemicals as the British and Dutch firms. The Netherlands more resembles the UK model (food and kindred products, bulk chemicals) than the German model (Philips electronics being an exception). France occupies an intermediate position (car manufacturing being an important employer). The same applies with respect to Italy (bulk chemicals like the UK and the Netherlands, also metal manufacturing and car production). Switzerland and Sweden resemble the German model, though with their own specialities as mentioned above.

The UK model is resource-based and consequently more internationally oriented from the point of view of production facilities in the early stages (e.g. in Third World countries). The German model can be labeled as 'Weiterverarbeitung': that is, concentration on the latter stages of production, mostly located in the home country, but with an international market orientation for durable goods. Historically these differences can to a high degree be explained by the different backgrounds of 'inward-looking' Central European nations versus 'outward-looking' Northwest European countries with a colonial tradition (UK, the Netherlands).

Functional Hierarchy and Geographical Concentration within Corporations

Economic-geographic theory states that the functional hierarchy of firms corresponds with specific spatial hierarchies as expressed in the center-periphery concept or in the notion of city systems.

Headquarters of corporate firms are supposed to be concentrated in the world's great metropolises, as tops of national urban hierarchies. These management centres control lower levels of these hierarchies and, more particularly, peripheral areas.

As is shown in Fig. 7.3, two 'models' can be discerned: the French case of superconcentration and the German case of deconcentration with geographical clusters at the lower spatial scale of capitals of Länder. UK firms are also concentrated to a high degree. Paris and London are consequently extreme examples of management centers. If *Dun and Bradstreet* figures are employed, it is found that 306 of the 500 top-management locations (of the largest European corporate firms) are concentrated in cities with at least four corporate seats (De Jong & Thuis 1983). London is at the top of this list (98 headquarters) followed by Paris (64) with an even higher rate of companies located in the capital. Other European metropolises are of far less importance.

The locational choice of top-management headquarters in the Federal Republic of Germany has strong historical roots (a political fragmentation during the period when most German firms were founded in the 1860s) and is partly resource-bound (the Ruhr region). As Monheim (1972) discovered, in the 1860s a change came about in the attractiveness of German management centers; Munich and Düsseldorf acquired a dominant position and pushed the formerly predominant centers of Hamburg, Frankfurt and Cologne to second rank. Although this cognitive element ('mental map') is a significant motivating force, the historical background and the international orientation of large companies (intercontinental airports!) also influence this locational pattern (the Ruhr region and Frankfurt consequently remain management centers). De Randstad is an example of a multilocational top-management center, while in Belgium, top management is concentrated in Brussels, a prime location for management

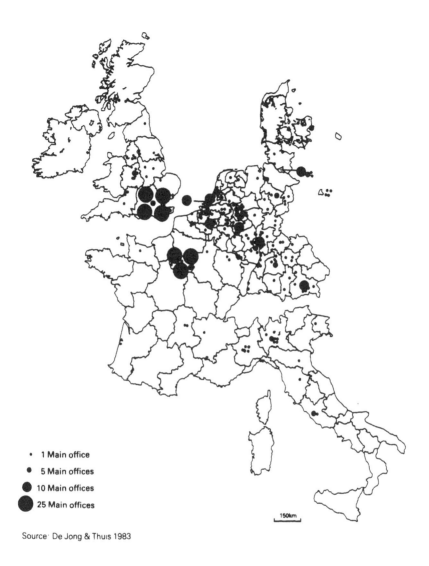

Source: De Jong & Thuis 1983

Fig. 7.3   Headquarters of the top 500 companies in
selected West European countries

within Europe.

London-based companies controlled 4.3 million jobs
worldwide in 1979 and Paris-based companies controlled 2.3
million out of 11 million jobs within the top-500 European
firms listed by *Dun and Bradstreet* (De Jong & Thuis 1983).
The Ruhr region, including Cologne, contains twenty-five
headquarters controlling 1.6 million jobs, somewhat more

than those controlled by firms located in De Randstad (nearly 1.3 million jobs).

Per sector, corporate firms may have a different hierarchy of functions based on locational production preferences, for instance resource-oriented petrochemical plants and steel mills or labor-oriented factories for textiles or electronics. These elements of Weber's locational theory still exert some influence!

As far as light manufacturing firms are concerned, at least half of the jobs are located within a radius of 100 km of the headquarters. The remaining jobs are mostly located in branch plants in peripheral areas within the country at a distance of 200 to 300 km from the top-management seat. The traditional core-periphery concept stresses the dependence of branch plants, which often leads to a tremendous loss of jobs and even closures of establishments in times of economic recession. Furthermore these branch plants are supposed to utilize low-qualified labor, and in the event of takeovers, they would lose skilled jobs. As Watts (1981) pointed out, such general ideas are much formulated too simply. In his dissertation, Keizer (1985) pointed out that very often such establishments in the Frisian area of the Netherlands acquire a broader range of competence (e.g. sales departments), assuming an 'intrapreneurial' character. This relates tes to the provision of management for other departments of the same firm, a consequence of the functional decentralization policies of large enterprises. Moreover, there is a structural change in the economy linked to the product life cycle, implying the collapse of old, highly specialized manufacturing areas. These areas were hard hit by the recent economic recession, much more so than peripheral regions of Europe, which are characterized by modern factories producing goods in earlier stages of the product life cycle.

Changes in the Direction of European Investment: The American Challenge Revisited

As Servan-Schreiber (1967) observed, European firms ran into difficulties in the face of the aggressive American way of operating on the frontiers of world markets. In the Western World nobody was aware in those days of an awakening Japanese challenge!

Ernest Mandel (1968) criticized the ideas of Servan-Schreiber. He agreed with the basic notion of a strong lead of US enterprises in the world arena, but his explanations were basically different. His opinion was that better marketing strategies and management attitudes were of secondary importance. The US lead in technology could be traced back to oligopolistic marketing structures fostered by mergers and acquisitions resulting in economies of scale.

This process of industrial concentration and capital accumulation was also criticized by John Kenneth Galbraith (1966) as the technological imperative of US business, deriving profit from technological spin-offs of defense contracts. Western Europe was divided in many respects. Around 1970 a wave of mergers and acquisitions occurred and multinationalization of European firms increased rapidly, even during the long economic recession of the 1970s and early 1980s.

During the economic boom (1960-1973) foreign direct investment throughout the world increased at the same rate as international trade, by as much as 50 percent more than the growth of the gross national product of industrialized nations (Smith 1985). During periods of recession, the growth of investment in the US by European-based firms was striking (OECD 1981). Of course there was good reason to expand in the US market, which showed better perspectives than the stagnating EC market. This strategy could, however, only be carried out on the basis of larger corporations through a process of mergers and acquisitions. At the same time there was a growing need to explore the rich US sources of innovation and to avoid potential trade barriers (McConnell 1983). This shift of European investment to other continents had a wider range than just the US. Newly industrialized countries also offered opportunities for expansion, as demonstrated by the fact that between 1975 and 1981 the (limited) growth of a sample of twenty-four European companies turned out to take place completely outside of Europe (De Jong & Thuis 1983).

As Vernon (1971) points out, the growth of the US stock of investments abroad, measured by re-investments and loans, was four times larger than the flows indicated in the 1950s and early 1960s. 'Multiplier' effects must consequently be taken into account, and the position of countries of origin may be different on the lists of stock and in the flow of direct investment. This can be illustrated for the Dutch situation. Corporate firms originating in the Netherlands hold the third position on the world's list of stock of investment abroad (behind the US and UK), comprising 10.5 percent in 1960, and 7.8 percent in 1980 (Japan, FRG and Switzerland being runners-up). If one looks at the flows of direct investment abroad, the Netherlands was far less important, comprising 2.6 percent of the world's total of these investments in the early 1960s and 4.9 percent at the end of the 1970s (Stopford & Dunning 1983). Traditionally three major companies (Shell, Unilever, Philips) were fully responsible for these international investments, but in recent years medium-sized enterprises have internationalized their activities, more in particular by moving into the US market (by acquisitions). The same trend has been traced with respect to Swedish company subsidiaries establishing in EC countries (Schröder *et al.* 1984).

An answer to the 'American Challenge' has been given. Twenty years later, the flows of international investment are more complex than ever before. European-based firms participate to a greater degree. In some sectors (e. g. computer hardware, aircraft) American supremacy continues, but in several other sectors (e. g. chemicals, electronics, machinery) the European position has been strengthened, fostered by a process of concentration. However, a new challenge has arisen in the meantime: Japanese expansion.

Big business, the core of modern capitalism, has survived a long economic recession and two energy crises. In addition to American giants in the world of big business, European as well as Japanese corporations have recently started to play a more important role. International investment has grown rapidly, even during the period of industrial restructuring. The UK and the Netherlands are the most multinationalized countries of Europe: their outward-economies are characterized by resource-oriented firms, partly in the traditional consumer products. The UK model contrasts with the German model of firms, being more engaged in durable consumer goods and capital goods. France and Italy contrast with Germany in the fact that they are to a high degree characterized by government control of big business, while their involvement in international business is far less than that of other European nations.

Management centers are concentrated in primary cities in the UK and even more so in France (e. g. London and Paris), while in Germany a deconcentrated pattern prevails. The traditional core-periphery concept has been questioned in view of a recent trend toward an 'intrapreneurial' strategy, which entails a decentralization of management tasks within corporations. However, restructuring regions have been dealt some severe blows. A tendency to locate subsidiaries outside Europe, in the US as well as in newly industrialized countries, must not be overlooked. European firms taking their investments abroad constitute the European answer to the American and Japanese challenges.

Multinationalization: The Position of the Netherlands

Up to this point, the focus has been on the large corporation. Corporations from all over the world were ranked by sales volume and employment, which may have suggested that each large corporation is a multinational. However, the degree of multinationalization, defined as the amount of investment abroad in relation to the sales volume, varies substantially among corporations. Referring back to the 'American Challenge' of Servan-Schreiber, it is clear that in the 1950s and 1960s, many European corporations lagged behind their American counterparts in their urge to invest. The immense size of the American domestic market created economies of scale for American corporations. In conjunction with the process of mergers and take-overs, this brought about the dominance of large corporations within an oligopolistic system, a market in which there are a limited number of suppliers working in league with each other. Mandel (1968) pointed out that this was one of the causes of the rapid accumulation of capital in American business life. This in turn brought about what John Kenneth Galbraith dubbed America's 'technological imperative'. He used this term to describe the nation's lead in innovations, which was generated by a high rate of expenditure on research and development, often in the context of defense contracts (entailing subsequent civilian spin-offs).

Twenty years ago, not a word was heard about a possible Japanese challenge or about the multinationalization of European business. In the period 1960-1973, at the crest of the wave of economic growth, foreign direct investment grew at the same rate as foreign trade and 50 percent faster than the GNP of the developed countries (Smith 1985). In the meantime, a number of changes took place. Initially, the major motivation for firms to go multinational was the comparative cost advantage. For American corporations, for instance, this justified investment in a European production plant to take advantage of the lower wage costs. Later, access to the growing and more integrated market of the European Community became the dominant reason for multinationalization.

The European counteroffensive was a second change. The process of concentration in British industry and the limited and stagnating domestic market in the UK induced British corporations to invest on the Continent and in the United States. Within the EC, large European corporations were already able to reap the benefits of the expanding market. The subsequent spate of mergers and take-overs provided extra fuel for further multinationalization, especially through investment in the United States (Knickerbocker 1971). Apart from market penetration, the aim was to set up research and development units and, even more emphatically, to acquire existing firms. This served several important objectives: to build upon the available innovation potential, to avoid the impact of protectionist measures and even to export from this new base of operations (McConnell 1983). Servan-Schreiber's words were heeded to the letter, and an answer to the American Challenge was found.

While the focus on the United States is clearly notice-able in the pattern of foreign direct investment emanating from each Western European country (OECD 1981), it is espe-cially pronounced in the flow of capital from the Nether-lands. In the mid-1960s, investment in the United States amounted to more than 1 percent of the total foreign invest-ment made by Dutch firms. By 1970, this had increased to 10 percent and in the past 10 years to about one-third of the total. The orientation toward investment in other member states of the European Community was consequently reduced from 70 to 45 percent.

Why did the outflow of capital from the Netherlands triple between 1972 and 1982? A number of factors seem to be responsible. The severe economic recession persisted for years and created overcapacity in the production sector and excessive wages in the assembly sector. The general invest-ment climate remained poor and, on top of this, the value of the Dutch guilder remained strong. Together, these factors precipitated the rapid multinationalization of Dutch corpo-rations (Fig. 7.4; De Mare 1985). But the flow of foreign investment entering the Netherlands also doubled in the 1970s, in spite of the decreasing attractiveness of the Netherlands in comparison to other countries (McKinsey 1978; cf. Fig. 7.3). One of the reasons for this relatively high level of activity during this period of economic recession is that businesses can be bought for relatively low prices at precisely such times.

The increase in investment abroad does not imply that the process itself was novel. For many years, the Nether-lands took third place among the ranks of foreign investors. It still maintains this position, even though the strength of its status has diminished somewhat (1960: 10.5 percent; 1980: 7.8 percent). Nevertheless, by 1980, the Netherlands was the largest foreign investor in the United States. Along with the Japanese and the Swiss, the Dutch share the dis-tinction of having a negative balance in the flow of foreign capital.

Consequences of Multinationalization for the Netherlands

Of the 200 largest manufacturing and trade firms in the Netherlands, 160 have been more or less thoroughly inter-nationalized. And of these, 98 are of Dutch and 62 of foreign origin. Half of the top forty Dutch enterprises book more of their sales abroad than in the Netherlands (De Jong 1985b). Has this expansion abroad been at the expense of employment in the Dutch establishments of these firms? There is indeed evidence to the affirmative.

In the case of the Netherlands, an outward-bound multi-national may be loosely defined as a Dutch firm with its own production and distribution facilities in various countries.

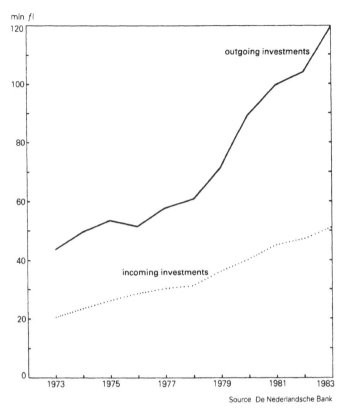

Source De Nederlandsche Bank

Fig. 7.4   The Netherlands: incoming and outgoing
international investments

Using this definition, during the latter part of the 1970s,
half of such multinationals reported a decline in employment
in the Netherlands and approximately half registered a
growth in employment in their foreign branches (De Mare
1985). These were for the most part the same firms. The
decline in employment amounted to a loss of 56,200 jobs,
whereas the number of jobs had stabilized somewhat in the
early 1970s. It may be inferred from the aggregate figures
that although the initial growth in foreign branches was
intrinsic, it was later due to the transfer of activities
from the Netherlands (Fig. 7.5). A similar dynamic was
observed among new multinationals that were formed since
1975. Their growth in the first half of the 1970s, an increase of 142,500 jobs in the Netherlands, was substantially
greater than the 60,500 jobs created in the latter part of
that decade when the firms expanded abroad (De Mare 1985).
These figures do not provide absolute proof of the inference
that growth abroad was at the expense of jobs at home,

147

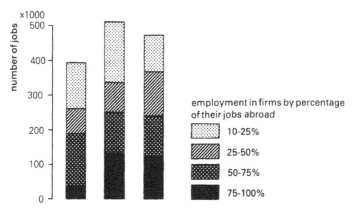

Source: De Mare 1985

Fig. 7.5   Employment in the Netherlands by Dutch
multinational corporations

because it is difficult to determine whether specific jobs
have indeed been siphoned off. In addition, the observed
decline in employment in the Netherlands is partially due to
the fact that the firms taken into consideration were primarily enterprises with an outdated product package. New
activities, in contrast, are undertaken relatively more
often in foreign establishments. However, the aggregate
figures are highly consistent with the explanation advanced
above.

The multinationalization of enterprise in the Netherlands has attained a wider scope. This is demonstrated by
the fact that in 1970, 92 percent of the jobs in foreign
establishments of outward-bound Dutch multinationals were in
the four biggest Dutch firms (Shell, Philips, AKZO, DSM),
whereas in 1980 these firms accounted for 'only' 79 percent
of the foreign employment provided by Dutch multinationals
(De Mare 1985). The increase in multinationalization was not
confined to manufacturing; the construction industry and the
large-scale retail sector took numerous initiatives in other
countries, which often involved take-overs.

The impact that the sixty-five biggest outward-bound
multinationals have had on Dutch industry may be summarized
in a few statistics. In 1980, these multinationals accounted
for 30 percent of the sales, value added and employment. At
the same time, they were responsible for 42 percent of the
export (even adjusted for the differentiation by industry
group, this figure is well above average). In the chemical
industry, the level of all indicators was even around 50
percent.

The above considerations leave no room for doubt that

the economy of the Netherlands is heavily involved in the process of multinationalization. The precedents for this tendency may be found in the very presence of head offices in the country and the continued existence through the years of many establishments of large firms. In addition, the exceptionally open economy provides a training ground for global activities. Yet the increased options to relocate production elsewhere puts more and more domestic industrial jobs in jeopardy. In this respect, the process of multinationalization has made the economy of the Netherlands more vulnerable.

7.3 Foreign Enterprises in the Netherlands

<u>Developmental Stages and Theoretical Explanation</u>

The background of the arrival of foreign enterprises forms a fascinating illustration of the way the Netherlands conformed to the West European industrial pattern. By keeping the cost of labor relatively low, the Netherlands had a comparative cost advantage in the 1950s. The small market, however, was an obstacle. Still, the effect of newly established foreign enterprises on employment was considerable. In the period 1950-1963, nearly 30,000 jobs emanated from direct foreign industrial investment, which amounted to 15 percent of the growth in industrial employment (De Smidt 1966). Despite a decreased governmental effort, this growth continued in the following years, favored by the economic boom. In the period 1964-1972, again nearly 30,000 jobs were created by direct investment in industry. Dutch enterprises suffered a massive drop in manufacturing employment, amounting to a loss of 83,400 jobs during the period 1964-1972 (Kemper & De Smidt 1980). The Netherlands had an initial advantage: its labor costs were lower than in neighboring countries. But around 1963, this advantage disappeared entirely. In the meantime, however, the formation of a Common Market had made good progress and the Netherlands reaped increasing benefits from its function as the 'Gateway to Europe'. This ushered in a strong economic growth of the sales market in the 1960s. Only during the long economic recession (1972-1983) did the number of new foreign establishments decline drastically.

Two approaches jostle for precedence in the explanation of the scope and the destination of direct international investment. The 'macro' explanation of Dunning (1980) is couched in the theory that the international division of labor is based on comparative cost differentials of immobile factors of production. This theory also accounts for economies of scale attained by specialization and vertical integration. Moreover, it accounts for the stage in the life cycle in which the various sectors are found at a given time. Allowance has also been made for the life cycle stage in the 'micro' approach of Kojima (1981); the business strategy is central to this approach. Oligopolistic market developments are facilitated by actions of large, financially strong enterprises that profit from their technical advantage and from economies of scale in production. On the basis of the principle of the product life cycle, these large enterprises send their production lines to suitable locations, taking account of the ensuing cost of the interaction. Rather than being incompatible, these two explanations are complementary.

Newbould (1979) and Hakanson (1979) make a useful differentiation of the process of internationalization in the following steps: export, hiring exclusive sales representatives, obtaining licenses, establishing in-house sales offices and, subsequently, service and distribution centers. The process may culminate in the decision to start a production company. In the 1950s, when comparative cost differentials (specifically cheaper labor) were manifest, it was often decided to establish a production plant elsewhere. In the 1960s, the transactional costs, which are incurred by transporting goods and introducing them on a market, made it attractive for a firm to have its own production facility close to the gateway to Europe. Yet the progressing market integration resulting from the development of the Common Market (EC) removed the necessity for European firms to establish a production company in a neighboring country. Only British firms would continue to establish firms on the Continent. This was not only due to the late entry of the United Kingdom into the Common Market; the permanent geographical advantages for transport and the higher productivity of labor formed additional reasons. Besides, there were no good investment opportunities in Britain at the time.

During the economic recession, the Netherlands had difficulty gaining the favor of foreign investors. West Germany was the biggest competitor. The cost of labor in the Netherlands had risen to the German level in the meantime. But for specialized products, the German labor market offered more highly trained personnel, particularly in the capital goods industry. The productivity of labor was also generally higher in Germany, its domestic market was larger and the international financial press considered its business climate to be better. Great Britain proved to exert a greater attraction than the Netherlands on American investors in particular, who found the UK a better alternative for labor-intensive, less highly valued production processes. Belgium offered good investment facilities (McKinsey

1978).

In the 1980s some recovery was booked in this competitive position through rigorous wage control, improvement of the business climate and an active recruitment policy, whereby, incidentally, many governmental agencies contended for the favor of foreign investors (Van der Poel & Valkenhoef 1985). Four problems are still pertinent to the attraction of foreign firms to the Netherlands from the United States and Japan in particular:

1 The market is too small (although the integration of European markets is in the offing);
2 Wage costs are high (despite recent wage controls);
3 Investment bonuses are lower than in Scotland and Ireland;
4 The Netherlands is not widely known as an industrial nation (Buck 1985).

Foreign Investment

At the end of the 1970s the proportion of total gross investment in the Netherlands that was due to direct intervention by foreign firms had returned to the level of the early 1960s. The economic recession had thus taken its toll (Van Nieuwkerk & Sparling 1985). In 1983 the invested capital from foreign sources (with at least a 20 percent share in enterprise) amounted to nearly 52 billion guilders, twice the amount recorded in 1973. This increase is scarcely spectacular, however, when corrected for inflation. The biggest investor, the United States, curtailed its European investment activity in response to the acute rise in the unit cost of products, the substantial decrease in economic growth and a decline in the rate of exchange for the dollar. In relative terms, the United States retained first place in the ranks of foreign investors in the Netherlands (1973: 37 percent; 1983: 35 percent). The members of the Common Market showed a considerable decrease in their propensity to invest; their share in foreign investment dropped from 43 percent in 1973 to 31. 5 percent in 1983 (West Germany was responsible for this change). Within the Common Market, the profit margins were much smaller than in the US, which was the destination of a great deal of European investment. Some compensation was offered by the newcomer Japan (1983: 2. 5 percent; in 1973, still less than 1 percent). Yet more was found in (a combination of) artificial constructions, such as the so-called Antilles route and Swiss investments, both of which have a tax-related background. In most cases, the compensation was found in Dutch firms that established holding companies in those countries.

The flow of investments changed course in another sense

(Van Nieuwkerk & Sparling 1985). In 1973, 73 percent of the investments were still in manufacturing; 10 years later, manufacturing accounted for only 57 percent. Trade, transport and other services are increasingly selected as destinations for foreign direct investment. The 'Gateway to Europe' function has undergone a change in emphasis. The role of Schiphol Airport has gained importance. Services are at the forefront in this shift. The importance of the port of Rotterdam as a location for industrial production has suffered from the effects of the oil crisis (refineries and primary petrochemical industry) and the decline of world trade. Within industry, especially investment from the United States and Great Britain in the sectors of energy, refineries and basic chemicals showed a strong decline.

The level of investment made strong headway in trade and in sectors such as banking and insurance. This demonstrates the importance of commercial centers at the gateway to Europe. It is especially American, British and Japanese investors that seek to penetrate the Continental side of the Common Market (sometimes limiting themselves, for example, to representation in Amsterdam's financial center).

The extent to which investments have had an impact on locations and employment deserves attention at this point.

Foreign Establishments

One out of every seven jobs in industry was paid for by a foreign firm; in total, this applied to 203,545 jobs in 1,981 establishments as of 1 January 1984, of which 55 percent of the jobs and 71 percent of the establishments were funded by direct investments (Loeve 1984). In particular, the British firms sank a great amount of capital into Dutch enterprises, due to a lack of expansion opportunities at home. There are some marked differences between acquisitions and direct investments (Table 7.4). The acquired establishments are larger (by a factor of 2) and they are also found in traditional industry groups.

In line with the theoretical notions offered above, it should come as no surprise that the impulse for direct investment is greatest outside the Common Market. Establishments that have been taken over are usually subjected to a reorganization, and definitely so during a recession. In the period 1980-1984, acquisitions by foreign firms cost Dutch industry 9 percent of its employment; in foreign-owned new industrial plants less than 4 percent of the jobs were lost. Dutch-owned/operated industry was confronted with a job loss of 12 percent, which is primarily explained by closures and cut-backs in the textile industry and shipbuilding (industry groups without foreign establishments). The picture is not noticeably different for trade and services (Loeve 1984). A

Table 7.4 Employment in foreign companies, by sector, 1984.

sector	greenfield establishments abs.	%	acquisitions abs.	%	joint ventures abs.	%	total abs.	%
manufacturing	64,447	58.1	48,078	59.4	7,277	61.8	119,802	58.9
trade/distribution	29,274	26.4	18,130	22.4	1,288	10.9	48,692	23.9
services	16,736	15.1	14,654	18.1	3,212	27.3	34,602	17.0
other	423	0.4	23	0.1	2	0.1	448	0.2
total	110,880	100%	80,885	100%	11,779	100%	203,544	100%

Source: Loeve 1984

third, modest category of foreign activity, the joint-venture, even shows a clear gain in employment. For firms from the United States and Japan, this form provides an opportunity to market technologically advanced and patented products via a company established specifically for this purpose. At the same time, this form allows foreign firms to take advantage of existing industrial premises, market channels and labor market contacts available to the well-established and often renowned Dutch firms.

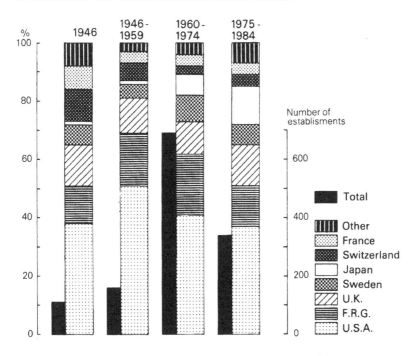

Source: Loeve 1986

Fig. 7.6 Greenfield plants of foreign companies in the Netherlands

153

Structural Changes in Foreign Direct Investment

In the long run, a number of structural changes have taken
place in the composition of foreign direct investment in the
Netherlands (Loeve 1984).
1.   By country of origin, the initial domination of the
United States has eroded (1941-1959: 51 percent; 1975-1984:
37 percent of the number of establishments; see Fig. 7.6).
American enterprises, as discussed above, established production plants in foreign countries early on. After the
growth spurt of the Common Market was over, there were very
few signs of new American interest. Japan is a late bloomer.
Japanese firms are making a powerful advance on the European
market, along with supporting sales establishments, but
their contribution to industrial investment is scanty.
British and Scandinavian firms seek access to the European
market through trade or industrial plants in the Netherlands. The background of British moves has already been
discussed above. In the past ten years, a number of small,
specialized Swedish firms have located in the Netherlands.
These firms can usually only afford one production company
within the European Economic Community. Sometimes, alongside
a production plant elsewhere in the Common Market, they use
the Netherlands as a springboard for export to third countries because of the excellent transport opportunities and
the 'international mindedness' found in the Netherlands
(Schröder *et al.* 1984, Loeve 1985).
2.   From the perspective of the industrial sectors, there
is a shift from manufacturing to trade and services. The
share of manufacturing establishments amounted to 38 percent
in the postwar period of industrial emancipation (1941-
1959); this share has dropped to 10 percent over the past 10
years. Swedish establishments remain more clearly oriented
toward manufacturing, while Japanese establishments are
primarily involved in trade. In terms of jobs, manufacturing
remains predominant, with around 60 percent in 1984. But
there too, trade and services show a strong growth (2,491
and 796 jobs, respectively, in 1980-1984, against a loss of
2,171 jobs in manufacturing).
3.   The contribution of foreign firms to the industrialization of the Netherlands is also qualitative in nature.
The Netherlands is historically weak in the so-called inter-
mediary sector which comprises firms that supply products to
other companies. This refers especially to capital goods
industries, such as tool and die making, in which at least
22 percent of the jobs counted in 1984 were provided by
direct foreign investment. Afterwards, during the 1960s, the
chemical industry came into the picture, where almost 18
percent of the employment was funded by direct foreign
investment. This share amounts to 8 percent for all indus-

Table 7.5 Employment in greenfield establishments of foreign companies.

industry	US abs.	US %	FRG abs.	FRG %	UK abs.	UK %	Sweden abs.	Sweden %	Japan abs.	Japan %	Other abs.	Other %	Total abs.	Total %
food and kindred products	1,637	2.8	-	-	765	6.3	-	-	-	-	411	2.5	2,723	2.5
textiles	540	0.9	99	0.9	25	0.2	380	3.6	-	-	266	1.6	1,310	1.2
lumber, paper	576	1.0	68	0.6	-	-	1,601	15.1	68	2.6	231	1.4	2,476	2.2
chemicals	11,516	19.6	1,806	16.8	4,346	40.7	350	3.3	49	1.9	2,361	14.1	20,447	18.5
metals	1,002	1.7	150	1.4	50	0.5	2,810	26.5	-	-	1,116	6.7	5,177	4.7
machinery	11,569	19.5	2,966	27.7	406	3.8	1,516	14.3	-	-	771	4.6	17,228	15.6
electronics	7,833	13.2	641	6.0	-	-	-	-	-	-	78	0.5	8,552	7.7
other	2,163	3.7	549	5.1	388	3.6	1,767	16.7	584	22.5	1,115	6.7	6,566	5.9
manufacturing	36,836	62.2	6,279	58.6	5,890	55.1	8,424	79.4	701	27.0	6,349	38.0	64,479	58.3
transport equipment	387	0.7	324	3.0	315	2.9	8	0.1	292	11.2	2,838	17.0	4,164	3.8
chemicals	1,085	1.8	434	4.0	968	9.1	207	2.0	-	-	736	4.4	3,430	3.1
machinery	1,731	2.9	871	8.1	352	3.3	622	5.9	211	8.1	457	2.7	4,244	3.8
electronics	8,145	13.8	580	5.4	449	4.2	636	6.0	465	17.9	500	3.0	10,775	9.8
other trade	3,244	5.5	1,161	10.8	537	5.0	311	2.9	246	9.5	1,162	7.0	6,661	6.0
trade	14,592	24.6	3,370	31.4	2,621	24.5	1,784	16.8	1,214	46.8	5,693	34.1	29,274	26.5
transport/commun.	908	1.5	733	6.8	410	3.8	97	0.9	128	4.9	635	3.8	2,911	2.6
financial	661	1.1	179	1.7	499	4.7	21	0.2	278	10.7	3,011	18.0	4,649	4.7
other services	6,209	10.5	161	1.5	1,265	11.8	279	2.6	275	10.6	1,007	6.0	9,196	8.3
services	7,778	13.1	1,973	10.0	2,174	20.3	397	3.7	681	26.2	4,653	27.9	16,756	15.2
TOTAL	59,206	100%	10,722	100%	10,685	100%	10,605	100%	2,596	100%	16,695	100%	110,509	100%

Source: Loeve 1984

tries and only about 4 percent for 'other' industries (after subtracting employment in machinery and chemical industry). The recent appearance of high-tech industry is noteworthy, although this development did lead to industrial production in American firms and not in Japanese enterprises. Table 7.5 demonstrates the international trends in specialization. Sweden has metal and machinery industries, as does Germany. The United States adds chemicals and electronics to these industries. The Japanese concentrate on electronics and automobiles, and the British specialize in chemicals.

Locational Pattern

During the Dutch emancipation in which it caught up with West European industrial trends (1952-1963), the contribution of direct foreign investments was characteristic in a regional sense as well (De Smidt 1966). When (re)locating production, the Dutch firms showed an increasing orientation toward the so-called problem regions in the North and South; the foreign firms offer a contrasting picture. There was a preference, certainly among American firms, for locations in or near the large cities of Western Europe, and this applied to the De Randstad as well (Hamilton 1976). Their 'mental map' did not correspond to that of the Dutch entrepreneurs, who did not have to contend with as much uncertainty in regard to the regional dispersal policy. Nor did British and Swedish investors have qualms about the effect of regional planning on their businesses, since they were accustomed to this type of policy in their own countries.

Nonetheless, the regional locational picture of the foreign companies gradually conformed to the locational pattern demonstrated by the Dutch firms, although De Randstad still played an important role. Of the current industrial employment in foreign firms established in the Netherlands in the period 1946-1954, 47 percent is concentrated in the three western provinces. For the period 1960-1974, this percentage is only 31. New preferential areas emerged in the 1960s, such as the large-scale port development, which is almost exclusively in foreign hands, along the Westerschelde Estuary. Another example is the ascendancy of Noord-Brabant with an (initially) large supply of labor and an abundance of industrial estates at a time when De Randstad was 'sold out' in this respect. Of increasing importance is that foreign firms seek entry into the rapidly growing and integrating European market. They tend to locate along the transport axes between De Randstad and the center of gravity of this market, which is shifting in a southeasterly direction within Europe (Kemper & De Smidt 1980). This leads to a growing tendency to avoid the North as a location for new business (Loeve 1985). A similar change of preferences is

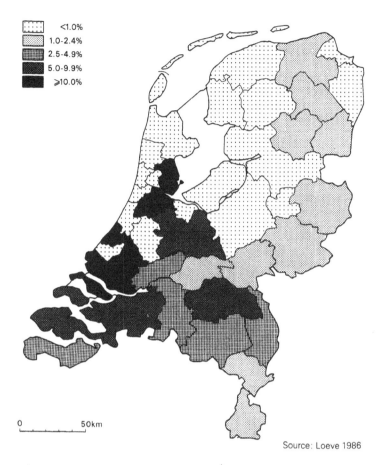

Fig. 7.7   Employment in greenfield manufacturing plants
in 1984 (the Netherlands = 100)

found among Swedish firms: firms with a traditionally strong
preference for establishment in the North of the Netherlands
would not repeat this choice if given the opportunity
(Schröder *et al.* 1984). New manufacturing establishments
tend to locate in Noord-Brabant. The clear overrepresentation of foreign manufacturing establishments in Greater
Amsterdam, Utrecht, Rijnmond, Zeeland and large parts of
Noord-Brabant is not reflected in the locational pattern of
recent initiatives. The north wing of De Randstad (including
Amsterdam) recently came clearly into the picture as well;
in addition, especially the West and the central part of
Noord-Brabant are favorite sites. The overcapacity in bulk
chemicals has forced the Golden Delta (including Rijnmond)
to drop out of the picture.

157

The share of foreign manufacturing establishments in the total regional industrial employment is highest in Zeeland and, despite the recent growth, much lower in Noord-Brabant (Fig. 7.7). The growth among existing foreign firms in the North masks the recent disinterest in locating there. In the northern wing of De Randstad, growth in existing establishments exceeds expectations; this is also the case in the southeastern part of the Netherlands (eastern Noord-Brabant and Nijmegen). In thirty out of forty areas, the development of employment in foreign establishments is favorable compared with the figures for industry in general. This is definitely the case in De Randstad and the southeastern part of the Netherlands. This confirms trends reported earlier. The commitment of foreign manufacturing

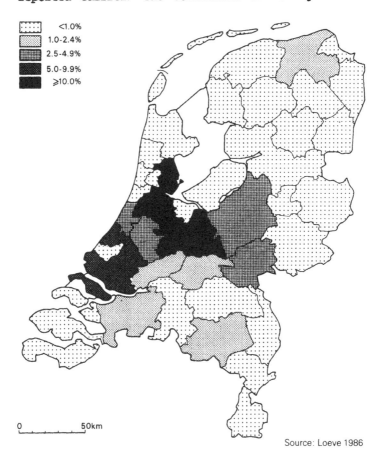

Source: Loeve 1986

Fig. 7.8   Employment in greenfield trade firms in 1984
(the Netherlands = 100)

158

establishments to the metropolitan environment is currently - with the exception of the north wing of Randstad - hardly stronger than that of Dutch establishments (Loeve 1986).

The position of the northern wing of De Randstad, close to Schiphol Airport and decision-making centers and with access to a highly qualified labor market, remains strong. This applies even more to foreign establishments involved in trade, of which almost 40 percent (accounting for 56 percent of the employment) are found in the northern wing of De Randstad. Within the metropolitan area of Amsterdam, an international orientation prevails (for example, Japanese firms with a strong orientation toward European markets; see Loeve 1985). Within the metropolitan area of Utrecht, the orientation is primarily toward the domestic market, according to an investigation by Bekkers *et al.* (1985). There is a slight tendency toward deconcentration further to the east. With respect to trade, one out of every three jobs is located in Greater Amsterdam, Greater Rijnmond and Utrecht (excluding bulk trade); for foreign establishments, this figure is two out of three (Fig. 7.8).

Although Greater Rijnmond hardly counts as a location for foreign trading firms, the three large city regions in Noord-Holland and Zuid-Holland almost monopolize the service sector, with four out of every five establishments (and even 86 percent of the employment) being located there. The central part of the Netherlands hardly plays a role in the service sector. The overemphasis on the large cities comes into relief when the specializations are incorporated in the picture. Financial services and those catering to airlines are located in Amsterdam, transport services are concentrated in Rotterdam and the offshore sector is found in The Hague. The picture almost looks like an 'over-exposure' of the old familiar geographic division of labor within the Netherlands (Fig. 7.9).

Foreign entrepreneurs are apparently not pioneers of a new locational pattern. Even more than the Dutch entrepreneurs, they select proven locations. Because the foreign firms are often highly specialized, they are well able to hold their own in the metropolitan environment. They also seem to be sensitive to the European market, as demonstrated by recent tendencies of newly locating industrial establishments in Noord-Brabant and trade and service establishments near Schiphol Airport. American and Japanese firms have their center of gravity in the northern wing of De Randstad; German firms are concentrated in the southern wing (in The Hague, where for example Siemens is established, and in the port areas of Rotterdam); although British firms tend to be concentrated in Rijnmond, they are also located in peripheral problem areas. Swedish firms in the trade and service

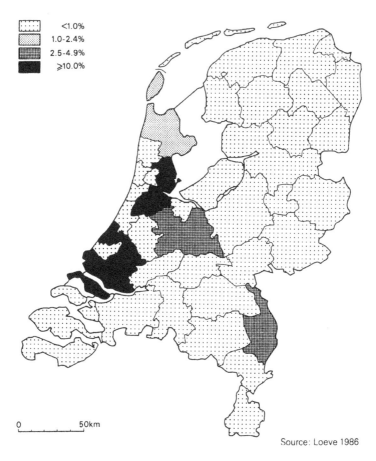

Source: Loeve 1986

Fig. 7.9   Employment in greenfield service firms in 1984
(the Netherlands = 100)

sectors are the only ones found in De Randstad. Their locational preference shifted from the North to the South of the
Netherlands due to their increased European market orientation.

7.4 The Netherlands: 'Gateway to Europe'

It was previously observed that the Netherlands forms an
attractive location for foreign firms. It was also noted
that Dutch industry exports relatively much unfinished
manufactures. In a response to this situation, the WRR
report entitled 'Place and Future of Dutch Industry' (1980)
advocated a qualitative up-grading of the industrial export
package. Still, neither the attraction the Netherlands holds
for foreign firms nor the orientation of Dutch industry

160

toward relatively low-value products are surprising. Both aspects are related to the gateway function that the Netherlands fulfills within Western Europe.

This gateway function is associated with the centrality of the Netherlands in terms of transportation within Western Europe. It is also related to the central position of Western Europe in the flow of commodities in world trade. This position is tied to the presence in Western Europe of the following:

- a large and affluent population;
- much export industry that works with foreign raw materials;
- an economic system that imposes relatively few restrictions on the international transport of goods.

The gateway function is most clearly manifested in the so-called hubs: places where goods and/or people arrive by some mode of transport and depart by another mode or continue their journey in another form. The two most important hubs in the Netherlands are the seaport of Rotterdam and Amsterdam's Schiphol Airport. Both are of international standing. Schiphol ranks sixth in the world in terms of the number of international passengers and tenth in terms of freight (Table 7.6). In terms of volume of goods handled, Rotterdam is even the largest port in the world (Fig. 7.10). The gateway function provides the Netherlands with comparative advantages. In general, the comparative advantages of a national economy are based on the competitive strength of the individual (manufacturing) firms, on specific characteristics of the production environment or on a combination of these two factors. Advantages related to the production environment may be permanent in nature. In contrast, there are also advantages that are temporary in nature (tax benefits, location subsidies).

Table 7.6 Rank of airports, by international arrivals and departures (x 1,000), excluding transit passengers, and by tons of freight (x 1,000) in 1982.

Passengers		Freight	
London/Heathrow	22,217	New York/J.F. Kennedy	1,009
New York/J.F. Kennedy	13,639	Chicago/O'Hare	720
Frankfurt/Rhein-Main	11,912	Los Angeles/International	652
Paris/Charles de Gaulle	11,281	Frankfurt/Rhein-Main	604
London/Gatwick	10,120	Miami/International	523
Amsterdam/Schiphol	9,666	Tokyo/Narita	510*
Hong Kong/Kai Tak	8,648	Paris/Charles de Gaulle	468
Miami/International	7,589	London/Heathrow	441
Zürich/Kloten	7,549	Atlanta/Hartsfield	339
Paris/Orly	7,324	Amsterdam/Schiphol	317

* Figures for 1981

Source: N.V. Schiphol

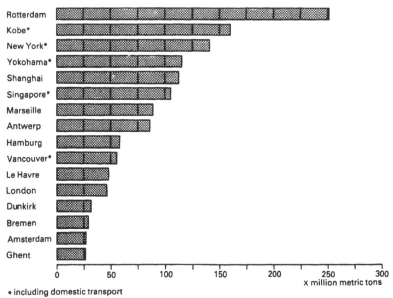

* including domestic transport

Source: Rotterdam Port Authority

Fig. 7.10 Sea traffic for a number of important harbours,
           1985

The gateway function is related to a rather permanent
element in the Dutch production environment: the centrality
within Western Europe. This may be seen as a strong and
stable aspect of the Dutch economy. Yet at the same time it
explains why the Dutch export pattern deviates from that of
countries such as Switzerland and Austria.

Many probably associate the expression 'The Netherlands: Gateway to Europe' with transhipment and hinterlands.
Transhipment is understood as transport of goods via an
airport or seaport, whereby both the place of origin and
destination of the goods are in foreign countries and the
items do not undergo any processing in the airport or
seaport. The transhipment links the gateway function of the
Netherlands to the primary or transport function of the
ports. An example is the iron ore and grain that arrives in
seagoing vessels and is then loaded (either at the dock or
riding at anchor) onto river freighters and subsequently
shipped to the hinterland. Another example is the promo-
tional campaign launched by Schiphol Airport under the
slogan 'Schiphol, London's Third Airport'. Potential
airline passengers in the vicinity of the English regional
airports were thus instructed that Schiphol would be an
ideal alternative to London as a place to initiate an
intercontinental flight.

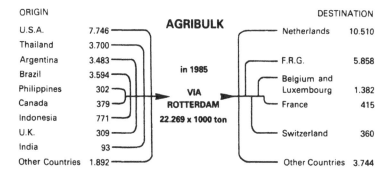

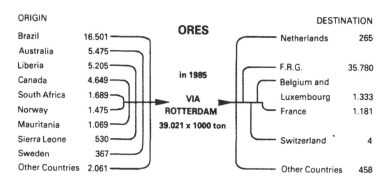

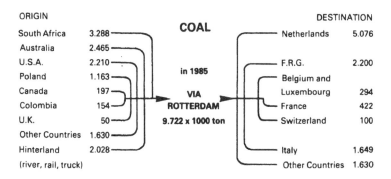

Source: Rotterdam Port Authority

Fig. 7.11 The hinterland of the port of Rotterdam for selected bulk products

163

The transhipment function undeniably occupies a key position in the gateway function of the Netherlands. A few characteristic examples referring to the seaport of Rotterdam are given in Fig. 7. 11. The transhipment of the bulk products listed there is not channeled to foreign destinations exclusively. These activities also involve foreign firms. Various oil companies, for example, take part in the transhipment of unrefined oil which is mostly conveyed through pipelines, the RRP (Rotterdam-Rijn Pipeline to the Ruhr area) and the RAPL (Rotterdam-Antwerp Pipeline). Besides the Dutch-British oil company Shell, participants in the RRP include the American companies Mobil Oil, Chevron, Texaco and the German company Veba Oil. The RAPL is used by Esso and Chevron as well as the Belgian company Petrofina and the British company BP. In Rotterdam, the companies active in the transhipment of coal and ores include EMO (Europees Massagoed Overslagbedrijf) and Ertsoverslagbedrijf Europoort. The participants in EMO are two Dutch firms (Frans Swarttouw and SHV-Steenkolen Handelsvereniging), two German firms (Thyssen and Ruhrkohle) and one French firm (Manufrance). Ertsoverslagbedrijf Europoort is owned and used exclusively by three German firms: Thyssen, Krupp and Mannesmann.

Despite the great importance of the transhipment function, the gateway role of the Netherlands is not limited to transhipment activity. A seaport that functions as a transportation hub also has secondary functions, such as storage, distribution, trade and industry. The difference between the primary transport function and the secondary industrial function is only one of degree. If iron ore arriving from Brazil with a destination in West Germany first undergoes simple processing in Rotterdam (removal of impurities) before being transported up the River Rhine, then there is more at hand than transhipment alone. In seaports and airports, countless simple processing activities take place which imply 'indirect transhipment' (Wever & Ter Hart 1986) and which are direct results of the gateway function. Moreover, every seaport and airport seeks to develop secondary functions alongside the primary activities. The reason is that these additional functions provide extra employment and a higher value added. This is especially true for the industrial function. For the Netherlands, the distribution function in particular is becoming increasingly important. To limit transport and storage costs, radical changes are being made in the transport of goods. The scope of these changes is illustrated by the situation at the ITT corporation, one of the biggest manufacturers of telecommunications equipment in the world (Fig. 7. 12). Until recently, every ITT establishment in Europe arranged its own incoming and outgoing transport.

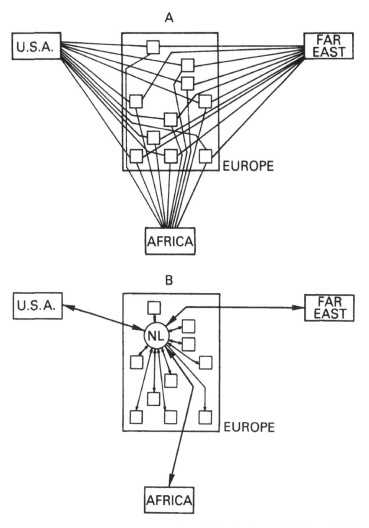

Source: Rotterdam Europoort Delta Journal 1983

Fig. 7.12 Supply and transit patterns of ITT-Europe,
before (a) and after (b) reorganization

Because the amounts involved were relatively small, the
costs incurred were high. A short time ago, ITT decided to
concentrate its activities. Rotterdam is now the central
storage and distribution center for all goods that are
shipped by water. For goods that are transported by air,
the main storage and distribution center is Schiphol
Airport. This concentration provides economies of scale

165

specifically for the intercontinental transport routes. This principle is also at the basis of the decision by the United States Line to focus its round-the-world service on the Delta Terminal (Maasvlakte) as the central European staging area for its largest container ships.

The gateway function of the Netherlands is of course not limited to its seaports and airports. Dispersed throughout the Netherlands, foreign firms manufacture their products with the aid of imported components. Some sell these products on the European market, others ship their products as components to other production units elsewhere in Europe or provide services to other plants. A case in point is the ITT corporation, but other examples are the plant of the American-owned Digital computer firm in Nijmegen and the establishment of the Japanese firm of Fuji in Tilburg. The full meaning of the concept of gateway function, however, can be grasped by describing the industrial developments in the Rotterdam port and around Schiphol Airport.

Rotterdam's Port Industries

Among incoming and outgoing goods in Rotterdam, oil continues to play an important role, even though the position of oil declined after the oil crisis. But Rotterdam is still the paramount oil supply port in Western Europe. Three factors in particular contributed to maintaining this status.

In the first place, this position is due to the steadily increasing size of the oil tankers throughout the 1950s and 1960s (economies of scale). The supply was therefore more and more concentrated in deep sea ports that were favorably located for surface transport. Secondly, it was enhanced by the pipelines laid since the end of the 1950s, which supplied unrefined oil to refineries located inland. To generate economies of scale, the participating oil companies laid large-gauge pipelines. Rotterdam became the origin of one of the most important of these, the Rotterdam-Rhine Pipeline to the Ruhr area.

From central ports, pipelines were then also laid to refineries in ports where larger tankers could no longer dock. Thus, the refineries of Mobil Oil in Amsterdam (which has been closed in the meantime) and Total near Vlissingen were connected to Rotterdam by way of a pipeline. These developments are illustrated by the changes in the position of Antwerp, for years an important oil port. Because the Westerschelde channel was too shallow, the Rotterdam-Antwerp Pipeline was laid in 1971; this pipeline currently supplies oil to the refineries located in the port area of Antwerp. When this pipeline was constructed, Antwerp became incorporated into the oil hinterland of Rotterdam. This partly explains why between 1970 and 1975 the amount of oil shipped to Rotterdam rose from 102.9 to 125.9 million tons; in Antwerp, the amounts landed fell from 26.8 to 8.2 million tons.

In the third place, Rotterdam has access to virtually ideal inland waterways connecting the port to its hinterland, of which the Rhine River is the most important.

Because of its position as an oil supply port, Rotterdam currently has five refineries. The oldest dates back to 1936 (Shell); the rest (Texaco, BP, Esso and KPC, the state oil company of Kuwait, which took over the interests of Gulf) were established in the 1950s and 1960s, when oil was one of the leading sectors in the Dutch economy. These refineries illustrate the port function in two ways. In the first place, they demonstrate the incoming and outgoing transport relations. The raw material is imported from abroad, apart from a very small amount drilled in the Netherlands (which is subsequently processed by Esso and Shell, the participants in the NAM, Nederlandse Aardolie-maatschappij). More than 50 percent of the product is also sold abroad. The refineries are thus located in Rotterdam, not so much for its proximity to the Dutch market but for access to other West European markets, a situation that has been observed among other international establishments as well. The port function is also manifest in the characteristics of the refineries located in Rotterdam. Refineries may be classified in simple and more complex types. The simple types cater to their 'own' regional market. They produce the standard products.' In addition to these, the complex refineries also make special products or specific grades. These special product lines are not narrowly targeted at their 'own' region markets but cater to other markets as well. Besides several simple refineries, the large oil companies also possess some complex types. Rotterdam has some complex refineries (Esso, Shell), which underscores the gateway function of the Netherlands.

The fact that Rotterdam has the function of an oil port is partially the result of the economic policies formulated after World War II. An extension of the secondary industrial port function was in line with these policies. This permitted a reduction in the nation's dependence on other countries (which is intrinsic to the port function) to some extent. The Netherlands had seen the disadvantages of dependency before 1940, when the German government adjusted its rail transport rates and improved its waterways in an attempt to route a greater amount of the goods destined for the Ruhr area through German ports such as Emden, Bremen and Hamburg.

In the oil sector, the attempt to industrialize was

successful. Although Rotterdam has built up a strong transhipment position in the handling of iron ore, the attempts to industrialize the processing of iron ore foundered there. Instead, the ironworks Hoogovens in IJmuiden took over this sector.

Based on oil, the secondary port function was expanded successfully. The refineries contributed to this success, but it was mostly based on the subsequent development of the petrochemical industry, which manufactures a wide array of products such as synthetic rubber, synthetic fibers, plastics, solvents, insecticides, raw material for paints and varnishes, and various kinds of fertilizer. The emergence of the petrochemical industry in Rotterdam was brought about by the ample supply of oil and refineries because of the gateway function for oil. All the raw materials of the petrochemical industry originate in the oil refineries. In addition, oil companies such as Esso, Shell and BP are among the most important producers of petro-chemicals. The outgrowth of the industrial activities in the port of Rotterdam has been aided by the establishment of the European market, which functioned as a catalyst. The location within the Common Market and the abundance of oil as a raw material made it possible for firms in Rotterdam to benefit optimally from the phenomenal growth of the market in the 1960s. This brought firms from non-EC countries to the port (Dow Chemical and Continental Columbian Carbon from the USA, ICI from the United Kingdom), as well as corporations from member countries, such as BASF, Hoechst, AKZO and DSM.

Because of the domination of oil and the strong links of the oil and the petrochemical industries, the Rotterdam port developed into an industrial complex based on oil, and ultimately on the gateway function of the Netherlands for Western Europe. It provided many multiplier effects for the Dutch economy through such industrial activities as ship repair, manufacturing of machinery, various assembly activities, maintenance, and traffic and transportation activities.

Also spatial effects became notable in the 1960s and early 1970s. Because of the threatening congestion, the government promoted a deconcentration of the port-related industries. The province of Zeeland, favorably situated midway between Rotterdam and Antwerp, has become the prime beneficiary of this overspill from Rotterdam. It led to the integration of the industrial complexes of Zeeland and Rotterdam. This is clearly manifested in the presence of the above-mentioned pipeline that connects the Total refinery in Vlissingen to Rotterdam. Likewise the petrochemical complex of Dow Chemical in Terneuzen is linked through pipelines to Vlissingen and to Rotterdam. In a later stage, a spatial effect was also felt in the Moerdijk area, some 25 km south of Rotterdam, where Shell built an entirely new facility when it lacked expansion space in its Pernis site in the Rotterdam port.

The oil and petrochemical activities of Rotterdam expanded at the time when it was the leading sector of the upswing of the Fourth Kondratieff. In the meantime, it has evolved to a more mature stage in its life cycle. Its growth rate is decreasing, in part because of political decisions, and increasingly the effects of competition are being felt. Especially the oil-producing countries in the Middle East and North Africa are expanding their production capacity rapidly. A second threat is embodied in the spatial shifts that are being registered in the world economy. With respect to manufacturing, the countries in the Atlantic region are losing ground to the growing economies in the Pacific theater. This is the outgrowth of the ascendancy of Japan and other industrial nations in Southeast Asia as well as the shift of the economic center of gravity within the United States, where California has gained in importance.

The influence of the shift from the Atlantic to the Pacific on the oil and chemical activities in Rotterdam is indirect. Because many products are imported from the Pacific region, the West European market for petrochemicals is shrinking. The imported television sets, VCRs, automobiles and ships contain vast amounts of plastics, synthetic rubber, paints, solvents, etc. Obviously, the dimmed perspectives for growth entail negative consequences for Rotterdam and for the Netherlands as a whole. This does not imply that the gateway function of Rotterdam will rapidly cease to exist. But attempts will be made to adapt to the changing international relations, which may temper the current emphasis on oil and basic chemicals and increase the importance of more sophisticated chemical products and other activities. Among such other activities, the distribution function will certainly be included.

Industrialization Around Schiphol

Measured by weight, the transportation of goods through the air amounts to only a fraction of the total volume moved by ocean-going vessels, even if bulk products (oil, ores, coal, grain) are not included. The average value of air freight, however, is 125 times higher per unit of weight. It is exactly this high value of the products which makes air transportation feasible. Consequently, air freight consists predominantly of expensive, high-quality goods (computer components, medical instruments, cameras, VCRs), of perishable goods which can only be transported over large distances by air (cut flowers, fresh produce, gramophone

Table 7.7 Weight composition of freight leaving Amsterdam/Schiphol
          airport on KLM airlines, April–October 1983.

1. flowers/plants	21%
2. machinery	13%
3. food and kindred products	10%
4. electrical appliances	7%
5. printed matter	6%
6. chemicals	5%
7. eggs	4%
8. livestock	3%
9. office machines	2%
10. aircraft parts	1%
11. photographic, optical, scientific precision instruments	1%
12. textiles	1%
13. precious metals	1%
14. other	25%

Source: KLM

records) and of wares, such as spare parts which have to be available speedily in emergencies (Table 7.7). For Schiphol, cut flowers and plants are of great importance; these products account for over 20 percent of the total volume of air freight by weight.

The nature of airfreighted products illuminates their most important difference from goods transported by sea. It could be argued that the characteristic bulk products of maritime transport reached their prime in the 1950s and 1960s, even though some of these (coal) have experienced a revival since then. They were the leading products in the upswing of the Fourth Kondratieff. In contrast, airfreighted products belong typically to the high-tech and/or high-touch categories. Because of this, they seem to conform to the upswing of the Fifth Kondratieff. This succinctly captures the essential differences between the recent and the expected growth of the 'Schiphol activities' on the one hand and the 'Rotterdam activities' on the other.

The gateway function of Schiphol for passenger traffic is best expressed in the number of transit passengers, those that connect to a different flight. Of all the passengers using Schiphol, 33 percent belong to this category (Fig. 7.13). This percentage probably makes Schiphol Europe's leading airport in this category. Its role in this sector is the result of deliberate policy. It was mentioned above that advertisements stressing this function of Schiphol were even circulated in England ('Schiphol, London's Third Airport').

Among the air freight, the proportion of transhipment is even higher, over 50 percent; this figure does not include the goods that are brought to Schiphol from abroad by truck or that are subsequently exported over the road. In recent years, the proportion of transhipment has grown

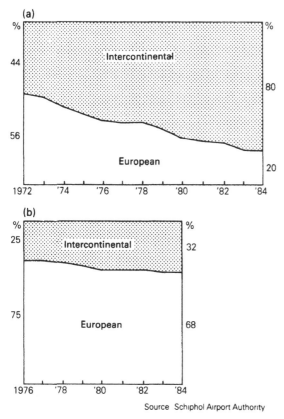

Fig. 7.13 Schiphol: transfer of goods (a) and passengers
(b)

continuously, which indicates further expansion of the
gateway function of Schiphol. This conclusion is also
supported by the evidence presented in Fig. 7.13a, which
expresses the growing importance of intercontinental air
freight.

The port function of Schiphol is particularly interesting because its 'natural advantages' were less than those at
play in Rotterdam. Certainly the central location within
Western Europe is also important for Schiphol, as this
densely populated and affluent area generates a high volume
of air traffic. Air transportation, however, does not depend
on physical amenities like the deep harbors or the Rhine
River which made Rotterdam the hub it is. The competitive
position of Schiphol among other airports is consequently
determined to some extent by completely different factors:
the *laissez-faire* trade policies of the Netherlands, the
pliant enforcement of customs regulations, the proximity of
warehouses, the presence of numerous transport companies,

171

good facilities for freight handling, the one-terminal lay-out, the fact that the airport is relatively quiet, etc. For freight transport, warehousing facilities are imperative. In 1964, Schiphol opened bonded warehouses and was the first European airport to do so. More than 60 foreign firms, especially Japanese and American, make use of these facilities. IBM, for example, selected Schiphol as its distribution center for Europe, the Middle East and Africa. Until recently, the twenty IBM establishments in Europe distributed their products themselves. Now all the establishments send their products to Schiphol, where the orders for the different countries are subsequently made up (see also Fig. 7. 12).

But besides distribution companies, Schiphol is an attractive location for other activities as well. Schiphol is the 'home' base of KLM, Royal Dutch Airlines. Fokker, the Netherlands' one internationally renowned aircraft manufacturer, is located there. Schiphol is also a prime location for companies that provide goods or services to firms at the airport (including KLM and Fokker), for companies that make intensive use of Schiphol in dispatching and receiving goods, and for companies that cater to Schiphol's passengers. This last category includes not only the tax-free shops but also the hotels that have been built around the airport.

These activities contribute to the strong spread of effects emanating from Schiphol which benefit the economy of the Amsterdam-Haarlem region in particular. The airport alone employs approximately 30,000 people. In addition, the surrounding area accommodates around 40,000 Schiphol-related jobs. Taking the multiplier effect into account, a great number of people are thus economically dependent on Schiphol Airport, directly or indirectly. It should be kept in mind that the airport provides highly skilled and well-paid jobs and that the activities at Schiphol and those in Schiphol-bound firms are basic industries in the national economy.

Schiphol is confronted with a structural shortage of space for business. This is a result, on the one hand, of the growth of air transportation activities and, on the other, of the restrictive policy exercised for many years by the province of Noord-Holland in regard to the residential environment. This is one reason why the area has experienced spatial spread effects since the beginning of the 1970s.

An increasing number of firms that have outgrown their premises at Schiphol tend to relocate outside the airport itself but remain in the vicinity. In a sense, Schiphol may be considered as a breeding ground for young firms. Numerous foreign firms with distribution activities for the West European market start out small at Schiphol. When the

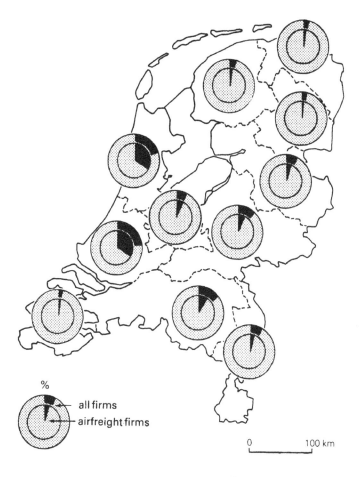

Source: Wever & Ter Hart 1986

Fig. 7.14 Location of firms using air transportation
(provinces)

establishment gets a foothold, expansion is considered,
conceivably supplemented with assembly activities. Because
of the scarcity of space, this entails a move away from
Schiphol for some firms. That is why IBM, Honeywell, Sony,
Hewlett Packard, Ricoh and Canon started out at Schiphol and
later fanned out into the vicinity. This deconcentration
process was actually promoted by the airport management.
Because of pressure on the available space, the airport gave
priority to so-called platform-related activities, just as
Rotterdam gave priority in the 1960s to activities tied to
deep-channel waterways. For activities that were not

173

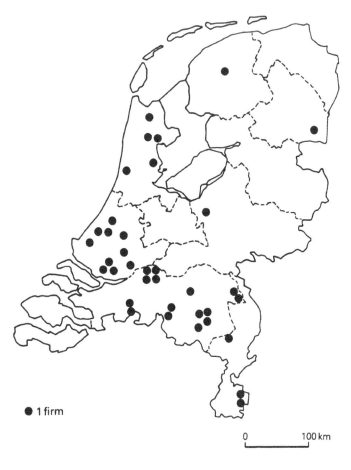

Source: Wever & Ter Hart 1986

Fig. 7.15 Location .of those firms using air
transportation at Schiphol and at another
airport

platform-related but were tied to the airport or sensitive
to its proximity, locations were sought close to Schiphol.
Fig. 7.14 shows that the immediate surroundings of Schiphol
especially benefited from these moves. In fact, 67 percent
of all 'air freight' firms in the Netherlands are located
in the provinces of Noord-Holland and Zuid-Holland (and
only 43 percent of all firms). The fact that the very
presence of an alternative can influence its use may be
inferred from Fig. 7.15. In addition to Schiphol Airport in
Amsterdam, firms located in the vicinity of Rotterdam,

Eindhoven or Maastricht also have recourse to the airports in these cities.

The perspectives for Schiphol's future seem to be good. Air transport and related activities are most likely to keep on growing. Schiphol has also built up a favorable market position in the course of time. Moreover, the general perception of the airport and the business activities that take place there has become quite positive. The role of Schiphol as a spearhead in the economy of the North Sea Canal area and, indirectly, of the Netherlands as a whole has been recognized and reinforced on all fronts.

8 Large Manufacturing Corporations in the Netherlands

8.1 Ten Large Manufacturing Corporations: Origin, Growth Strategies, Restructuring and International Orientation

Between the 1950s and the 1980s, the list of the top ten Dutch industrial corporations changed somewhat. Nevertheless, in terms of the number of their employees, the firms listed in Table 8.1 were continuously at the top of the list. Still, there are substantial differences among these firms with respect to their origin, their growth trajectory and the strategies they pursued, their international orientation with respect to the location of establishments, and the role of the Dutch branch plants within the corporation. The differences in their locational patterns and the specific regional development of the corporations is the topic of Section 8.2. In that section, the top ten are compared with the top twenty-five industrial corporations, and even, in as far as regional distribution is concerned, with the top 100 firms in the Netherlands. But, first, the focus will be on the characteristics of the firms themselves.

Origin

At first glance, the origin of the top ten industrial corporations falls into two groups. Some large corporations were created through mergers of existing firms (external growth), and others grew from humble beginnings and developed independently into large firms (internal growth, Jansen 1972). Early mergers characterized the two corporations which are by far the largest among the Dutch 'internationals': Shell and Unilever. In 1907, the 'Koninklijke Nederlandsche Maatschappij tot Exploitatie van Petroleumbronnen in Nederlandsch-Indie' located in The Hague ('Royal Dutch') and the Shell Transport and Trading Company in London joined forces (the distribution of shares being 60:40). The motive for the merger was to be able to compete with the, then gigantic, American corporation Standard Oil (its later break-up resulted in the present Exxon Corporation and other firms). In 1929, on the eve of the Depression, Unilever was engendered by a consortium comprising a group of British corporations led by Lever Brothers Ltd. and a group of firms of Dutch origin, N. V. Margarine Unie (which consisted of the margerine factories of Van den Berg and Jurgens, the Hartog slaughterhouses, and the oil and animal fat refinery Calvé). Also in this case, the merger brought together a number of corporations that dealt in raw materials - partly derived from the colonies - that faced worldwide competition from American firms.

This history sets Shell and Unilever apart from other industrial corporations that were created much later, around 1970, by mergers: the chemical corporation AKZO, the machinery firm VMF-Stork and the shipbuilding firm RSV. Only Dutch firms were involved in these mergers and the motivations were mostly defensive; this explains why they coincided with drastic reorganizations. In 1969, 'Koninklijke Zout-Organon' (which itself resulted from a union of chemical corporations) joined forces with the 'Algemene Kunstzijde Unie' to form the AKZO corporation. A group of expansive chemical firms were thereby joined with a company whose production of synthetic yarns and fibers was stagnating but who could nevertheless provide its chemical partners with much-needed capital. In fact, the form selected for the AKZO corporation was a conglomerate of divisions. A similar structure can be discerned in the two defensive mergers that brought about VMF-Stork and RSV. Both corporations form part of the capital goods sector, where reorganizations became necessary in the 1960s and took on dramatic proportions in the 1970s.

The 'Verenigde Machinefabrieken Stork' (VMF-Stork) was created in 1954 by the merger of Werkspoor (which included the railway coach manufacturer Rolma) and Stork, and it immediately became the biggest Dutch producer of heavy machinery. Throughout the 1960s, the firm remained the biggest corporation in the entire metal sector. The Rijn-Schelde- Verolme group (RSV) was only brought together in 1971 under strong pressure from the Dutch government in order to deal with the structural problems in the shipbuilding industry. A previous take-over by Verolme of the Amsterdam shipyards of NDSM in 1968 was a combination of a rescue operation and government support for a mammoth dry dock built for Cornelis Verolme in Rotterdam. Partner Rijn-Schelde was itself the offspring of a merger of the 'Koninklijke Maatschappij de Schelde' (in Vlissingen) and a number of shipyards and machine factories in Rijnmond (Wilton-Feijenoord, RDM).

Other Dutch industrial corporations grew as independent entities, though not always without government support. An example of such a corporation is the chemical corporation DSM, which grew in 1967 out of Staatsmijnen (incorporated in

1902). The basis for the change-over from coalmining to chemicals was laid in the 1930s when Staatsmijnen started some carbochemical activities. The state became a minority shareholder in the ironworks 'Koninklijke Nederlandse Hoogovens en Staalfabrieken' (Hoogovens). Its founding in 1918 was inspired by the strategy of creating a national basic industry. As was described in Chapter 2, the government provided major financial support for the foundation of this steel industry. The outgrowth of Fokker to the national aircraft manufacturer was at least partly inspired by the government's insistence on taking over Avio-Diepen (in Ypenburg) and De Schelde (in Dordrecht) in 1954 and Aviolanda (in Papendrecht). Another example of an expanding industry that fitted in with the post-war industrialization strategy was the development of the automobile industry. The core of this industry was formed by the construction company Gebr. Van Doorne (DAF) in 1928. In 1966, the State bought a 25 percent interest in order to help finance the establishment of an automobile plant in Zuid-Limburg, which was to play a role in the restructuring of this coalmining region. In 1978, the corporation was broken up into DAF-Trucks and an automobile manufacturing company, Volvo Cars. Of this company, the Swedish Volvo corporation owns a 30 percent share, while the Dutch state participates for 70 percent. A corporation that grew entirely on its own strength is the international corporation Philips Gloeilampenfabrieken, which was incorporated in 1912 as the continuation of Philips & Co., which was founded in 1892.

Direction of Growth and Restructuring

The Dutch industrial corporations differ with respect to their scale and their international orientation; these differences are manifest in the directions they take in periods of growth and reorganization. This sets the 'big three' (Shell, Unilever, Philips) apart from the others. In addition, there tends to be a correlation between external growth and the tendency towards diversification, which is generally based on arguments related to reorganization and spreading of risk.

The tendency toward diversification emerged during the 1960s. An authority in the field of corporate strategy (Ansoff 1965) insisted that corporations should establish a clear line in their policies. This was translated into the selection of a product-market scope. The most far-reaching policy would be the decision to diversify the corporation to include products and markets that were entirely new to the company. In reality, diversification could lead to the inclusion of adjacent product lines, but occasionally it leads to the combination of a wide range of activities that are a far cry from the original focus of the firm (the so-called conglomerates). Among the three large Dutch 'internationals', Philips and Shell are characterized by internal growth. Philips has broadened its range of activities over the years. Besides consumer products, Philips' establishments in the Netherlands manufacture 'professional' products for business, defense, health care, etc. (semi-conductors, telecommunications, medical instruments, etc.). Only the 1970 take-over of the firm Nederlandse Kabelfabrieken brought an external element into the corporation. Because this manufacturer of wire remained an incompatible element, it was partly sold in 1985. Shell took advantage of the expansion in the oil and chemical sector, but lost ground during the energy crises of 1973 and 1979. Within the Netherlands, Shell participates in other firms in the energy sector, such as 'Nederlandse Aardolie Maatschappij' (NAM; this corporation is a joint venture with Esso which, like Shell, owns 50 percent) and in Gasunie (Shell and Esso each own 25 percent of this corporation). For the international activities, the 1970 take-over of Billiton was of great importance. It allowed Shell to diversify by adding ore extraction and non-ferrous metal manufacturing to its activities. The acquisition of Billiton is an anomaly in the internal-growth strategy of Shell. (The corporation does own a 58 percent share of Wavin, a manufacturer of plastic products, but this activity is closely linked with the core activities of the oil company.) Unlike Philips and Shell, Unilever did expand its activities in the Netherlands by take-overs in the 1950s and 1960s. Its acquisitions were canneries and preserving factories (De Betuwe, Lucas Aardenburg), frozen foods (Iglo) and meat packing firms (Olba, Udema, Zwanenberg).

In the 1970s, the near absence of corporate take-overs of other firms located in the Netherlands is striking. This is the period in which most corporations cut back their labor force. Even the three 'internationals' conform to this pattern. But reorganization is much more typical of a number of national firms, especially of RSV and VMF-Stork. Table 8.1 illustrates this vividly.

The downfall of the RSV corporation was brought about by the global crisis in the shipbuilding industry. In most western industrialized countries, the coming of this crisis had been presaged by stiffer price competition in the 1960s. The situation deteriorated in the 1970s as a consequence of stagnating world trade and because of the energy crisis. Yet necessary reorganizations were postponed for years, which cost the government hundreds of millions of guilders in subsidies. The situation was only aggravated by speculative attempts at diversification and the top-heavy company hierarchies, and eventually many plants had to be closed. A few healthy companies were rescued in the nick of time.

VMF-Stork survived a series of reorganizations and is now repaying the debts it incurred with the government. This corporation expanded in the 1950s and 1960s by an active take-over strategy. The corporation's aim was to achieve a broader product mix in the capital goods sector (including such activities as the design and production of pipeline systems, fertilizer plants, air-filtration devices, cranes, and diesel engines). Also, this corporation was severely affected by the economic downturn and the energy crisis in the 1970s. The corporate management, however, failed to react in time to these changing conditions. When VMF-Stork eventually closed some of its plants, it was confronted with opposition among its employees; there was severe labor unrest in the case of the closure of the railway coach works Rolma in Utrecht in 1970. But the actions stopped short of an occupation of the plant by the workers, the type of protest that focused attention on AKZO's attempts to close down its Enka-Breda yarn factory in 1972.

Likewise other national corporations had to implement a reorganization. Akzo had combined all its activities in the field of fibers and yarns in its ENKA-Glanzstoff division, which faced a steep decline in the demand for its products after 1970. The corporation had organized itself into a number of divisions, each of which consisted of one or more of the leading firms acquired in earlier take-over activities. This structure makes it difficult to classify the development of this firm in terms of internal and external growth.

Basic industry, including Hoogovens and DSM, has also suffered from the long-lasting economic recession which led to a structural overcapacity. The diversification within the Netherlands remained limited. Yet Hoogovens entered the field of aluminum production (Aldel) in the 1960s and initiated various activities in the area of techniques for environmental control during the 1970s. Its metalworking division entered a long drawn-out reorganization after 1970, to which its subsidiary DEMKA eventually fell victim. DSM had participated in the reorganization of the entire coal-based industrial complex in 1965-1973. Its diversification is expressed in participations in Gasunie (40 percent), the former automobile manufacturer DAF (25 percent) and several firms located in Zuid-Limburg. Most of these activities were instigated by the government. Its participation in 'Chemische Industrie Rijnmond' fitted more closely in its original profile, as did the outgrowth of its fertilizer division.

The dominant strategy of the 1960s, of expansion and diversification, created complex and sluggish organizations. The disadvantages became clear during the recession of the

1970s. Consequently, the dominant strategy of the 1980s is decentralization of management ('companies within the corporation') and the sale of 'exotic' elements ('back to the basics', 'stick to the knitting').

Internationalization and Specialization

The Netherlands may be a home base for the three 'internationals', but even before the Second World War, these companies were building a global network of production facilities. Shell and Unilever maintain head offices in London in addition to their headquarters in The Hague, respectively Rotterdam. The Dutch share of the total employment offered by the two firms amounts to 'only' 15 percent for Shell and 4 percent for Unilever. But the quality of the jobs should not be underestimated. For example, in 1985, the international headquarters of Shell in The Hague employed 3,865, mostly highly qualified, people. The work force of its (internationally operating) laboratories in Amsterdam and Rijswijk numbered 2,377 persons, again for the most part highly qualified. In comparison with the Dutch division (Shell Nederland, with its head office in Rotterdam, its refineries and its chemical plants), which has a total work force of 8,120, the 'added value' of the international operations stands out clearly. The same applies to Unilever and Philips. In addition, with respect to their growth potential, supporting activities seem to be in a better position than production. Rationalization and automation exact their greatest toll among production jobs. With 21 percent of its employees located in the Netherlands, Philips has a far greater domestic work force than either of the other two 'internationals' and is by far the largest private employer in the country (Tables 8.1 and 8.2).

As far as their contribution to employment in the Netherlands is concerned, some of the 'sub-top' corporations make a much greater contribution than Shell or Unilever. Hoogovens is the second largest employer in the country, after Philips. With respect to its Dutch work force, AKZO employs as many persons as Shell (including Billiton). Like DSM, the company makes a more important contribution to employment in the Netherlands than Unilever (Table 8.1).

With respect to internationalization, AKZO, VMF-Stork and DSM, out of all the runners-up among the large corporations, have moved the greatest distance; they are strongly oriented toward the United States. AKZO could build on the international tradition that already existed within the ENKA-Glanzstoff division (a part of this division originated in Germany). In the 1960s, DSM had already set up facilities to procure input materials for its fiber division and its fertilizers in the United Kingdom and the United States.

Table 8.1 Number of jobs in ten large Dutch manufacturing companies, 1960-1985.

	1960	1971	1978	1985	1960-71 %	1971-78 %	1978-85	Percentage of jobs located in The Netherlands			
								1960	1971	1978	1985
Philips[a]	67,707	91,574	78,213	71,100	+ 35	− 15	− 9	36	27	22	21
DSM	42,100	19,500	24,082	16,605	− 54	+ 23	− 31	100	95	74	62
AKZO	19,256b	30,779	25,735	23,100	+ 60	− 16	− 10	40	34	30	36
Shell[c]	18,123	18,209	16,404	17,561	0	− 10	+ 7	9	10	11	15
VMF Stork	15,838	20,117	14,523	8,115	+ 26	− 27	− 44	93	88	73	68
Hoogovens[d]	12,400	23,331	26,100	27,383	+ 88	+ 12	+ 5	100	100	34	100
Unilever	10,859	16,715e	16,680	11,250	+ 54	0	− 33	4	5	4	4
Fokker[f]	5,150	6,786	7,237	9,640	+ 32	+ 7	+ 33	100	35	41	100
RSVg	−	26,800	22,923	−	−	− 14	−	−	90	81	−
DAF/	4,673	9,577	11,549h	5,991	+105	+ 21	+ 3	96	83	80	69
Volvo	−	−	−	5,886							89

a Excluding NKF.
b Employees of companies later incorporated in AKZO.
c Excluding Billiton.
d In 1978 part of Estel.
e Excluding Zwanenberg.
f In 1971 and 1978 part of Fokker-VFW.
g Constituent parts could not be calculated for 1960.
h In 1978, separate companies, no specific data available.

Source: Jansen et al. 1979, supplemented with data from Fin. Dagblad (1985)

Table 8.2   The twenty-five largest manufacturing companies in the Netherlands by employment (1985).

1. Philips (electro)	71,100	14. T & D Verblifa (packing)	4,598
2. Hoogovens (steel)	27,383	15. KNP (paper)	4,418
3. AKZO (chemicals)	23,100	16. Océ-v.d.Grinten (office equipm.)	4,135
4. Shell (oil, chemicals)[a]	22,000	17. Gist-Brocades (pharmaceuticals,	
5. DSM (chemicals)	16,605	biochemicals)	3,877
6. Unilever (food)[b]	11,250	18. Douwe Egberts (beverages)	3,740
7. Fokker (aircraft)	9,640	19. DMV-Campina (dairy)	3,397
8. VMF-Stork (machinery)	8,115	20. Melkunie-Holland (dairy)	3,397
9. BP-Nederland (oil)	7,928	21. Noord-Nederland (dairy)	3,116
10. Heineken (alcoholic		22. Suiker Unie (sugar)	3,100
beverages)	6,142	23. Dow Chemical Ned. (chemicals)	2,956
11. DAF-Trucks (trucks)	5,991	24. Coberco (dairy)	2,925
12. IBM-Nederland (computers)	5,904	25. CSM (sugar)	2,874
13. Volvo Car (cars)	5,886		

[a] Including Billiton.
[b] Data 1984.

Source: Derived from Fin. Dagblad, 5 September, 1986

An entirely different strategy with respect to inter-nationalization was revealed in the integration of the Dutch corporations Fokker and Hoogovens with, respectively, the 'Vereinigte Flugtechnische Werke' (Fokker-VFW) and Hoesch (the resulting corporation was Estel), both from Germany. Already in 1952, Hoogovens had acquired an interest in the 'Dortmund Hörder Hütten Union' (DHHU), which was expanded into a controlling participation in 1959. The cooperation of a blast-furnace at a coastal site, working with cheap imported ores, with a high-quality steel mill located at the center of a large market ('Weiterverarbeitung') seemed so attractive in 1972 that Hoogovens and Hoesch decided on a complete international integration with parity interests.

The merger of Fokker and VFW in 1969 was looked upon as the first step toward a European aircraft industry. However, there was no follow-up. The intended merger with Messer-schmidt in 1977 was cancelled. The basis for cooperation disappeared during the prolonged economic recession. At that time, the aircraft industry proved to be especially vulner-able because of its sensitivity to high energy prices. Consequently, the two corporations were separated in the early 1980s. Estel was also dissolved in that period, when it became clear that the huge losses suffered by the steel industry in general were especially concentrated in the steelworks of Hoesch in the Ruhr region. Divestment became imperative when the two corporations (and the two coun-tries!) could not agree on a formula to share the costs of a thorough reorganization. Consequently, the main office in Nijmegen, located halfway between the two plants, was closed down.

In recent years, internationalization has proved to be

identical with the trend toward acquisition of firms in the United States, a country selected for its positive economic outlook (cf. Section 7.2). European mergers involving large corporations have failed during the economic crisis.

8.2 Geography of the Top Ten and the Head Offices of the Top 100

Recent Position of the Top Ten

As was argued above, the domestic importance of the ten largest industrial corporations should not only be measured in terms of the size of its labor force within the Netherlands. But even in terms of employment alone, the ten corporations retain a dominant position. In 1985, they still occupied the top eight places on the list (Table 8.2; cf. Table 8.1). Two of the top ten had dropped off the list: RSV has ceased to exist and DAF-Volvo has been separated. The blue-ribbon league is followed by a set of large corporations that either form parts of 'internationals' (e.g. the oil company BP, the Heineken brewery and distillery conglomerate, and IBM) or consist of firms that emerged from large-scale reorganizations (e.g. the metallurgical firms T & D-Verblifa and KNP). The corporations in this class form a rather diverse group. They include internationally successful firms in 'high-tech' industries, such as the photocopier manufacturer Océ van der Grinten and the biotechnology firm Gist-Brocades, which do not owe their prominent position to their large work force alone. More remarkable is the presence of corporations in the foods and tobacco sector at somewhat lower ranks among the top twenty-five. Mergers have propelled new dairy product firms onto the top twenty-five list. A number of corporations in this industry with a strong international orientation (and high sales figures, such as Nutricia and ccFriesland) fall just short of being included. Obviously, employment figures do not tell the entire story.

Regional Distribution of the Top Ten and of the Head Offices of the Top 100

Over the years, the corporations have undergone a substantial (functional) decentralization and (geographical) deconcentration. The hierarchical organization of a corporation is assumed to be reflected in the spatial distribution of its activities. According to the theory of Törnqvist (1970), contact-sensitive and highly skilled work is performed at or near the headquarters of the corporation. The preferred location for such a main office is supposedly in a metropolis or another center offering specialized services. As the distance from the headquarters increases, the activities carried out in an establishment presumably make fewer demands on the qualifications of the labor force. Such remote branch plants would therefore employ mostly unskilled labor. If the process of deconcentration (internal growth) occurs because of a lack of unskilled labor in the central regions, this distribution would be further enhanced. When firms in the periphery are subject to take-over (external growth), a leakage of skilled personnel to the main seat of the corporation or to other central locations would result. In the Netherlands, such a process was discerned as a consequence of the VMF-Stork developments in the 1960s (Jansen 1972), but in recent years the leakage seems to have decreased in importance. According to Keizer (1985), establishments that have been acquired by another firm very often retain more highly skilled functions than in the past.

In the long term, the evolving organizational structure of corporations brings about regional differences in the quality of the employment offered. The labor market, however, brings together supply and demand. Regional differences in the supply of labor have always existed and persist even today, and corporations tend to react to these differences. This can be shown clearly for the corporations with a main seat outside De Randstad. Their establishments in De Randstad do not fit the model of less-skilled employment. On the contrary, the quality of the De Randstad branches is often above the corporate average. According to Jansen *et al.* (1979), this can only be explained by the nature of the production environment (labor market, informal contacts with government offices, presence of research centers, etc.).

The organizational structure of corporations that maintain their own central and peripheral establishments does not necessarily correspond to the generally accepted geographical notions of center and periphery, which can be illustrated for AKZO and Philips. But the geographical reality cannot be completely ignored, not even by the largest corporations. The selection of the location of the head offices of divisions of the corporation reflects the influence of their geographical structure. But the largest corporations are capable of creating a production environment outside a metropolis that is nevertheless suitable for even the most highly specialized activities.

A separate category in this sense is formed by the corporations that are (or were) tied to a specific location, such as a coastal site (Hoogovens, Shell Nederland, RSV, certain divisions of VMF-Stork and AKZO) or near the source of their raw materials (DSM, other divisions of AKZO). Such a specialized site may be located within De Randstad, although it is sometimes in a remote area (DSM).

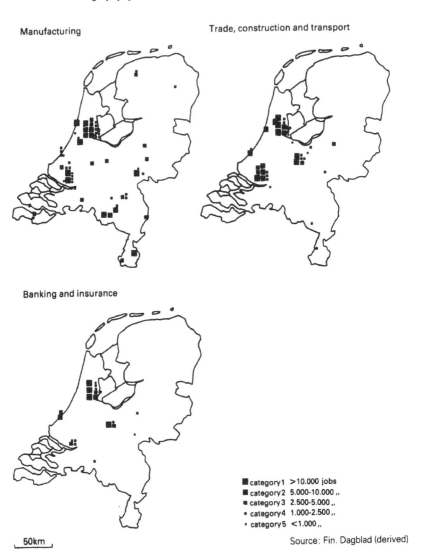

Manufacturing

Trade, construction and transport

Banking and insurance

- category 1 > 10.000 jobs
- category 2 5.000-10.000 „
- category 3 2.500-5.000 „
- category 4 1.000-2.500 „
- category 5 < 1.000 „

50km

Source: Fin. Dagblad (derived)

Fig. 8.1   Location of the headquarters of the top 100
companies and the top ten banks and insurance
companies (by sales and employment,
respectively)

A striking feature of the geographical distribution of
the top ten industrial corporations is the dispersal of
their headquarters. When compared with the pattern of the

Table 8.3 Regional distribution of main offices of the top 100* manufacturing,
construction, trade and transport companies, and the top twenty**
banks and insurance companies in 1985.

	Manufacturing	Construction, trade and transport	Banking, Insurance
Four main cities (suburbs included)	28	28	17
West (remaining centres)	12	4	-
Intermediate areas (Overijssel, Gelderland, Noord-Brabant)	14	4	3
Periphery (North, Zeeland, Limburg)	8	2	-
	62	38	20

* Rank order according to sales.
** Rank order of banks (10) according to assets and of insurance companies (10)
according to premiums collected. State-owned corporations not included.

Source: Derived from Fin. Dagblad, 5 September, 1986

top 100 firms in the manufacturing, construction, trade and
transport sectors, a clear contrast shows up between
manufacturing corporations on the one hand and non-manufacturing firms on the other (Table 8.3 and Fig. 8.1). The head
offices of 65 percent of the manufacturing firms are located
in the western part of the country, a substantially lower
share than among the non-manufacturing firms (84 percent in
the West). The intermediate regions (from Deventer to Breda)
are of major importance in the pattern of industrial
locations, also for firms other than AKZO, Philips and
Volvo. Dairy companies are a major factor in the periphery.
The dominant position of De Randstad is only reflected in
the distribution of purely commercial services, such as
financial institutions and insurance companies (Ter Hart
1979).

During the 1950s and 1960s, a clear expansion took
place in areas away from those where the ten largest
corporations have their headquarters. The corporation with
the largest work force in the Netherlands, Philips,
implemented its own regional policy in the period 1945-1973.
It deliberately established branch plants in the problem and
(later) incentive areas. Unilever also embarked on a process
of deconcentration in the 1960s, but mostly in connection
with its take-over strategy. Within the VMF-Stork corporation, similar tendencies could be discerned. In the 1970s,
however, deconcentration lost most of its earlier impetus.
Since then, structural selectivity has been shown to change
the regional distribution.

The impact of the ten largest firms on total industrial
employment is substantial, as demonstrated by the increase
in the share of jobs they provided: from some 20 percent in
1960 to over 25 percent in 1978 (Jansen et al. 1979). In the

North and the East, their impact was considerably lower than elsewhere. Especially in the South their impact was substantial. However, the process of deconcentration brought numerical (not qualitative) improvement for the northern Netherlands. In the 1960s, at least half of the increase in industrial jobs in the North was provided by the ten largest industrial corporations. Even if the jobs in firms that were taken over are subtracted, still one out of every three new industrial jobs added during this period of economic growth were provided by the large corporations. Their record in the South was negatively affected by the closure of the mines during the 1960s, which decreased the number of jobs at DSM. Similarly, the problems of RSV and VMF-Werkspoor during the 1970s caused the number of jobs provided by the ten large corporations to decline in the West. During this period, the regional effects of the restructuring of economic sectors became clear, not only outside the traditional problem areas but also in the western part of the country. Some compensation for the employment situation in the large corporations in De Randstad was provided by Philips. Even though this corporation maintains its headquarters in Eindhoven, it does provide highly skilled jobs in De Randstad as well as in Twente (Hengelo), Apeldoorn, Nijmegen and even in Groningen. The process of tertiarization, which coincides with an increase in the level of employment, is common to many establishments of the large corporations. It even occurs in their branches in the problem areas, but to a lesser extent and at a later period.

8.3 Summary and Conclusions

The list of the top ten industrial corporations has not changed much during the past 25 years. After their stormy development in the 1960s, often entailing mergers and take-overs, the lean years of the 1970s brought cases of drastic reorganization. In spite of this, the ten largest corporations remain in control of a quarter of the total industrial employment in the Netherlands. But they are no longer as eager to diversify as before. Hoogovens and Fokker retracted their strategy of linking up with large foreign corporations. The reorganizations that brought the downfall of RSV and the decline of VMF-Stork are typical of the heavy capital goods industry. The three large internationals weathered the crisis, but to both Shell and Unilever the operations in the Netherlands may be important in a qualitative sense, though limited in extent. Philips remains by far the most important industrial employer in the Netherlands. De Randstad is the most important center for the provision of services to the ten largest industrial corporations.

9 Regional Development Potential

9.1 Image and Reality: Regional Differentiation on the Wane

Previous chapters evaluated the spatial developments in Dutch manufacturing. This review provides a composite picture of reality that is at odds with conventional wisdom. The stereotype image of regional differentiation conceives of the West, or alternatively, De Randstad, as the location of the most highly specialized industrial activities. This is assumed to be the area where head offices and R&D activities are concentrated. In addition, the conditions for the renewal of the economy (product and process innovation) are considered to be favorable there. The popular consensus depicts the West as the region with the greatest economic potential in the Netherlands.

The antithesis of the economic core area is the peripheral problem region. This is generally viewed as a region that has to cope with low-value production activities that have been displaced from the West. These activities are presumed to take place in branch plants, which implies that the economy of the periphery is dependent on the central region (external control). Obviously, there is no room for R&D under such conditions, and whenever the economy turns down, many branch plants are closed down. The economy of these regions is not considered to be buoyant. Consequently, the recent trend in regional policy, i. e. the tendency to diminish the incentives to invest in the problem regions while emphasizing the intrinsic potential of each region, is commonly interpreted as 'betting on the strong' and as the abandonment of the weaker regions.

The developments described in the preceding chapters, however, show the conventional wisdom to be somewhat inaccurate. When short-term fluctuations are ignored, it becomes clear that the industrial development of the so-called intermediate zone (especially in Noord-Brabant) has been particularly strong. Consequently, the classical center-periphery contrast is no longer discernible. The pattern of successful industrial activity, the relative contributions of the regions to innovation, the participation in various innovation programs, the locations of head offices of industrial firms, the geographical distribution of large Dutch corporations, etc., all illustrate the narrow range of regional differentiation. If regional differences show up at all, the intermediate zone tends to occupy the best position, along with De Randstad. Obviously, there must be some explanation for this better-than-anticipated performance of the regions outside the West.

The fact that reality deviates from the stereotype is not because conventional wisdom is completely unfounded. The image of the economic performance of regions is based on processes that have occurred in the past or still do occur. However, they are characteristic of a differentiation at a scale which goes beyond that of the Netherlands. It incorporates elements of the discussion on the center-periphery pattern in the relationships of developed and developing countries. Myrdal's theory of cumulative causation (cf. Chapters 4 and 5) provides a case in point. Once popularized, the theory was applied to developments in North America and Western Europe. As an explanation of the current industrial pattern in the Netherlands, however, this cumulative causation theory is inappropriate. The stereotype also reflects the spatial interpretation of the product life cycle as explained in Chapter 3. This interpretation is derived from the experience in the United States, and, especially because of the differences in scale, comparisons between the United States and the Netherlands are often invalid.

Regional Economic Potentials

Why should explanations that are applicable to North America be invalid for the Netherlands? In order to answer this question, the popular but ill-defined concept of 'regional-economic potential' (Roelofs & Wever 1985) will first have to be explained. The concept entails at least three dimensions: the regional production environment, the regional production structure, and the 'quality' of regional business (which includes the organizational capacity of the region).

As was argued above, regional variations with respect to the 'quality' of industrial business are not very large. But considering the actual developments and the eagerness for innovation, the intermediate zone seems to be paramount in terms of 'quality'.

With respect to the regional production structure, a concentration of weak industrial sectors still typifies the image of more or less peripherally located regions. However, it was shown above that restructuring in such areas has proceeded significantly. The conventional image of the

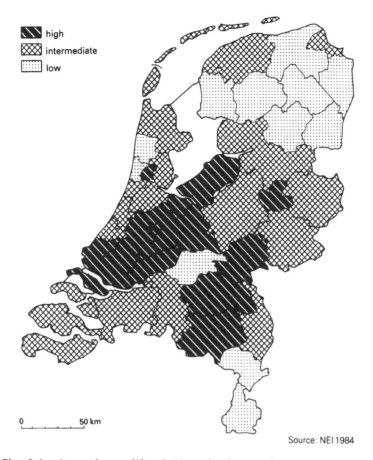

Source: NEI 1984

Fig. 9.1   Innovation profile of the regional economic
            structure

regional industrial specializations was based on the old
mosaic of functions in the Netherlands. Whereas these functions were once clearly differentiated, their boundaries
tend to become blurred as the regions become increasingly
similar. The distribution of successful activities illustrates that regional differentiation is on the wane, and
this trend has been confirmed in an investigation carried
out by the Netherlands Economic Institute (TRANSFER). This
study determined the innovation profile of the (largely
industrial) regional production structures on the basis of
various indicators (Fig. 9.1).

The production environment, the third dimension, comprises all those qualities of the region outside a firm's
control that influence the operation of business (infrastructure, industrial estates, educational institutions,

regulations and ordinances, etc.). Economic geographers tend to stress its influence because it varies from one place to another. The peripheral eastern part of the province of Groningen is in an entirely different situation than Rijnmond. Each region has some advantages. One area is therefore more suitable for certain activities than another. The spatial variation, however, is different for each locational factor. Some of these are permanent and unique, such as deep channels for navigation. Others are ubiquitous and subject to change. Industrially zoned land is available virtually everywhere, and more can be added rapidly. Industrial estates were lacking, in both quality and number, in some cities in the West for a long time, while in other parts of the country, overcapacity pervaded. Also at the lower end of the scale, especially with respect to cities and their surroundings, quality differences persist.

Apart from the issue of the spatial variation, the significance of the production environment for a firm remains an open question. Can a firm's profitability indeed be higher in one location than in another? The answer to this question is different in this modern age from what it would have been in 1935. Obviously, shipyards will always be compelled to locate on the waterside. Even now, a shortage of labor can induce a firm to consider a partial or complete move. On the other hand, over time the Netherlands seems to have become smaller. Even though some differences persist, nearly everywhere local facilities are of such high quality that the profitability of firms is determined more by its own 'quality' and its market perspectives than by the elements of its production environment. Regulations apply irrespective of location, wage levels are set through industry-wide collective bargaining, the provision of information and the activities of organizations are not influenced by location, etc. In addition, the construction of expressways has been a major factor in the convergence of production environments. This made it possible for firms that lacked certain elements in their regional production environment to fill this gap at little expense by utilizing the amenities of nearby complementary regions.

However, this does not apply equally to all elements of the production environment. A vivid illustration of this is provided in the above-mentioned study by the Netherlands Economic Institute. This report identifies potential for innovation of regional production environments by indicators such as: national and international accessibility (ports, airports), the presence of centers of learning, economies of agglomeration, availability of a highly skilled work force and/or of suitable business premises, an attractive residential environment, etc. Obviously, centers of learning, large cities and Schiphol Airport are within reach of every firm in the Netherlands, even if the establishment is not adjacent to a university or institute of technology, in a large city or close to Schiphol. But because of their inclination to avoid risk, firms do consider the distance to such facilities in their deliberations over the selection of a new location. Entrepreneurial behavior is not purely rational! With respect to the characteristics of the labor market and the residential environment, real quality differences do exist and are recognized as such by entrepreneurs. These differences will be elaborated on below.

Nevertheless, numerous industrial high-tech firms are located in production environments that are inappropriate by convential standards. The fact that they are there, however, does not imply that the production environment has become irrelevant; rather, it reflects the ease with which the necessary inputs can be brought in from elsewhere. This demonstrates the diminished influence of the regional production environment, which is in contrast to the persisting relevance of the factors of regional production structure and the 'quality' of the local business community. To paraphrase a remark by Allen Pred (1977), with respect to industrial activities, the Netherlands may be considered to have become one single 'urban field'. Obviously, this is not the same as one agglomeration in a morphological sense. Even sparsely populated areas as De Veluwe and the green heart of De Randstad are part of the Dutch urban field. It is in this sense that the Netherlands differs in scale from regions the size of North America and Western Europe, where urban fields alternate with areas harboring vastly different production environments. In the context of the Netherlands, the term regional-economic potential implies that nowadays the presence of capable entrepreneurs is more important than the quality of the production environment. Obviously, this was also the case in former days, but then entrepreneurs were more often faced with the necessity to relocate elsewhere. It is precisely this difference that has made the modern development of many regions much more positive than that of the past, which still lingers on in the popular image. The continuing improvement of the production environment of the so-called problem regions will further bolster this trend toward convergence.

Differences in Regional Production Environments: Labor and Housing Markets

Discussions of regional development potentials tend to focus on the map of regional differences in unemployment (Fig. 9.2). This map used to serve as the key element in the selection of incentive areas. At first sight, the unemployment problems are greater in the periphery than in De Rand-

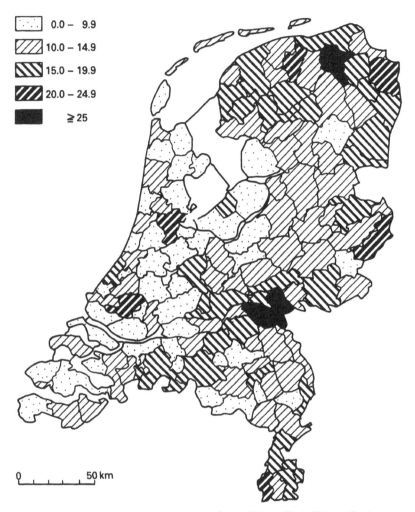

Source: Ministry of Social Affairs and Employment

Fig. 9.2   Unemployment in the Netherlands, as a
            percentage of the wage-dependent labor force
            (1986)

stad. Even among the regions that make up the successful
intermediate zone, major differences in the unemployment
rate are noticeable. How can this be reconciled with the
tendency of production environments to converge and the
tendency toward deconcentration in the national economy?

    Closer inspection shows that the unemployment rate in
the large cities comprising De Randstad is as high as in the

periphery and that, measured in absolute numbers, the employment situation is even worse in De Randstad than elsewhere. Cities are harder hit by unemployment than the countryside, while suburban areas are the least affected. This is indicative of the existence of a selection process: the cities are populated by increasingly large numbers of disadvantaged persons. At the same time, the local employment opportunities bypass the city's labor force, as the jobs are filled by people from the suburbs and rural areas (Van den Berg 1986, De Smidt et al. 1986). Hence, the unemployment problem in the cities is entirely different from that of the so-called restructuring areas (Zuid-Limburg, Twente, the eastern part of Groningen, Tilburg, Nijmegen), where the demise of industrial sectors aggravates the unemployment situation. Several rural areas where a large share of the labor force is involved in the construction sector also suffer from high unemployement rates (Fig. 9.2).

Is a region's unemployment rate indicative of its lack of opportunities for development? However paradoxical this may seem, the answer is that the two are hardly related at all. The adaptation of restructuring areas to new activities takes a whole generation. Re-training of older workers proves to be very difficult. But new, successful activities are being found in these areas, providing opportunities for a new generation of workers. In most regions, the unemployment problem is due to an oversupply of labor. Demand has not declined since the period of economic recovery started. Instead, the current unemployment problem is due to demographics: large numbers of young people are entering the labor force (Geerlof 1983). Obviously, in some areas there is more demand for labor than in others. This sometimes causes sharp contrasts to emerge in the spatial patterns. Within large city-regions, the line that separates city and surroundings coincides with the differentiation of the occupational skills of the local population; qualifications tend to be low in the city and high in the suburbs. Of the entire Dutch labor force, 14.1 percent is unskilled, but only 6.2 percent of the jobs that are offered through the labor exchanges require unskilled workers. In the cities, the supply of unskilled labor - partly composed of the so-called guest workers and their children - is even substantially larger. Their high unemployment rate is maintained by the decreasing availablity of opportunities for unskilled workers.

In determining the regional development opportunities, the economic structure of entire city regions should be taken into account. This indicator does not specify the chance of success in finding employment for various population groups, differentiated by level of training or place of residence. At the level of the municipality, local develop-

size of flows

balance

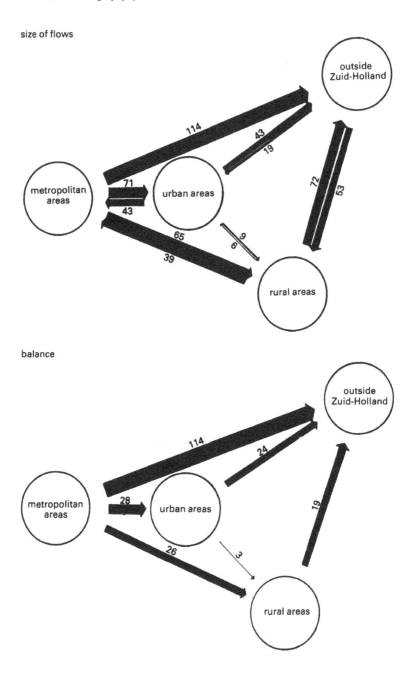

Fig. 9.3   The relocation of firms in Zuid-Holland, annual
average for the period 1976–1985

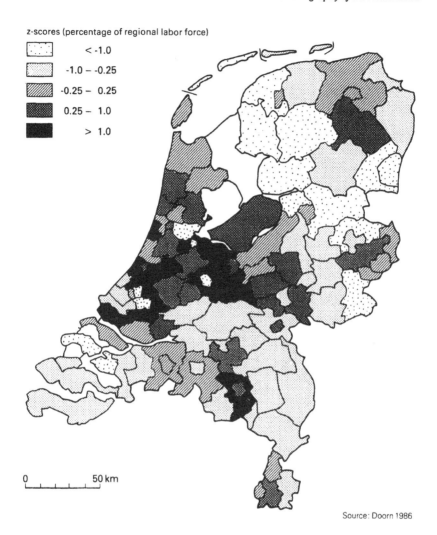

z-scores (percentage of regional labor force)

	< -1.0
	-1.0 – -0.25
	-0.25 – 0.25
	0.25 – 1.0
	> 1.0

0   50 km

Source: Doorn 1986

Fig. 9.4a The social map of the Netherlands (1981):
         regional distribution of men with higher and
         intermediate-level qualifications

ment opportunities are strongly related to the mobility
(suburbanization) of businesses. In the aggregate, there is
a continuous outflow of firms from the cities to the suburbs, and sometimes to other regions, for instance those at
the edge of the West (Fig. 9.3). In determining the existing
pattern of development potential, the present structure of

197

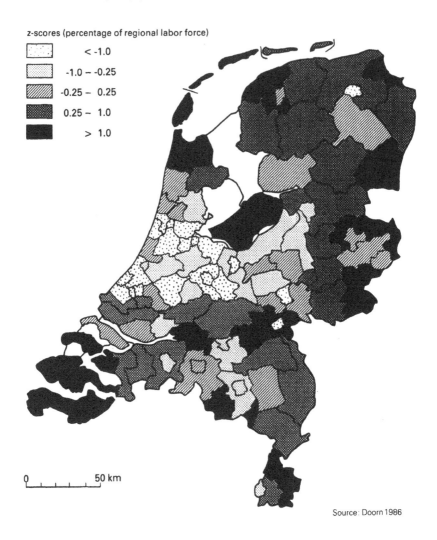

z-scores (percentage of regional labor force)

	< -1.0
	-1.0 – -0.25
	-0.25 – 0.25
	0.25 – 1.0
	> 1.0

0   50 km

Source: Doorn 1986

Fig. 9.4b The social map of the Netherlands (1981):
regional distribution of skilled male labour

the labor force should be taken into account, as well as regional differences in the level of training of the younger population.

The social map of the Netherlands (Fig. 9.4 a-c) shows a remarkable differentiation (cf. Doorn 1985 for more details). People with higher and intermediate-level qualifications are clearly overrepresented in De Randstad, especially

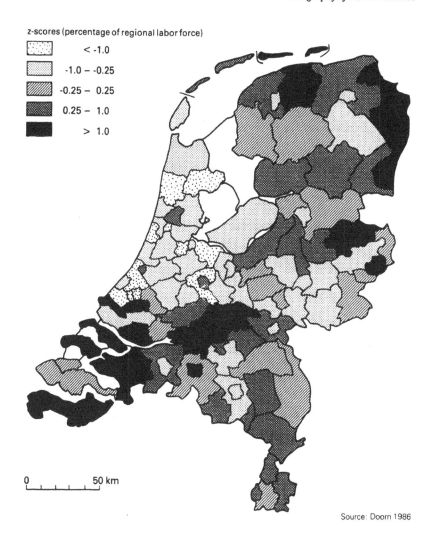

z-scores (percentage of regional labor force)

	< -1.0
	-1.0 – -0.25
	-0.25 – 0.25
	0.25 – 1.0
	> 1.0

0   50 km

Source: Doorn 1986

Fig. 9.4c The social map of the Netherlands (1981):
regional distribution of unskilled male labour

because of their prevalence in the suburbs and in De Veluwe.
Elsewhere, such a degree of overrepresentation is only found
in and near the cities of Eindhoven, Groningen and Maastricht. The distribution of the skilled workers is the exact
inverse of the pattern displayed by employees in management
positions. The unskilled are overrepresented in a number of
areas, but they fit the general pattern evinced by the

skilled work force. The areas with a relatively large inci-
dence of unskilled workers are mostly rural, but unskilled
workers are also overrepresented in several older industrial
cities: Rotterdam, Zaanstad, Tilburg and even in Utrecht.
Much of the rural area in the northeastern part of the
country shows an overrepresentation of the self-employed.
This group has not been mapped. Obviously, the patterns
sketched here are merely rough approximations of the regional differences in the qualifications found within the
(male) labor force.

The regional differences in the educational level of
younger persons is only relevant with respect to the lower
and middle rungs of the social ladder. People with higher
qualifications tend to be geographically mobile, and therefore the regional differences are of little consequence for
the development potential of the regions. However, attractive places of residence do have an advantage where migration is concerned. The suburban zones around the large cities,
the region of De Veluwe and the sandy areas in the province
of Noord-Brabant are clear favorites. This residential
preference map will be of increasing significance to the
location decisions of firms with a highly qualified work-
force. In the West, the growth centers have been built to
attract the lower and middle-level work force. These centers
therefore tend to draw firms that are being displaced for
lack of space in the older cities. Medium-sized cities have
an edge over larger cities in terms of the quality of their
residential environment (Van den Berg 1986).

Regional differences in the educational level are
particularly significant for industrial firms if they express differences in vocational schooling. There is a clear
trend toward convergence of regions in this respect, even
though this does not necessarily apply to all fields of
specialization. In spite of the overall tendency, a chemical
plant may still suffer from a lack of specialists among the
graduates of the regional school system.

Multinationalization: The Position of the Netherlands

Two caveats are in order with respect to the declining
significance of differences among the regional production
environments within the Netherlands. First of all, the
discussion applies to firms that are already located some-
where. If a firm happens to be located in the periphery,
there are now fewer incentives to move to the West than
previously. But the situation is entirely different for
firms that definitely want to move or have to do so. The
same applies to foreign firms that want to establish themselves in the Netherlands. In the selection of a new location, the characteristics of the production environment are

Source: Pellenbarg & Meester 1984

Fig. 9.5   The average assessment of 70 locations by Dutch
            entrepreneurs

are important, as was elucidated in Chapter 7. The production environment is not a factor for the sole reason that it
exerts a direct influence on the profitability of the firm.
Rather, its importance is due to many subjective factors,
most of which pertain to the production environment, that
play a role in the choice. Rightly or wrongly, perceptions
and evaluations of the various production environments
influence the locational choice and, indirectly, the economic potential of regions. The preferences expressed by
foreign firms have been discussed in Chapter 7. Fig. 9.5
presents the evaluation as expressed by Dutch entrepreneurs.
Their preference for the West and the central part of the
country (including Noord-Brabant) parallels that of foreign
decision-makers.

A second caveat with respect to the interpretation of
the declining significance of regional differences pertains
to the very open character of the Dutch economy. Like the
agricultural sector, Dutch industry is strongly export

oriented. One of the consistently strong points of the Dutch economy is its function as the 'Gateway to Europe'. Therefore, international developments, especially those that change the balance of the incentives to invest in certain countries, are of vital importance to the Dutch economy. Such international developments, which may have a severe impact on the economic potential of the Dutch regions, are now occurring. The economic center of gravity of Western Europe, the continent which forms the focus of the Dutch economy, is slowly shifting in a southerly and southeasterly direction. This is mirrored in developments taking place in individual countries. In England, the traditional industrial areas are losing ground, while London and the Southeast are gaining. In France, the zone Bordeau-Grenoble/Nice is growing rapidly in comparison to the Alsace-Lorraine and the region Nord/Pas de Calais. In the Federal Republic of Germany, Baden-Württemberg and Bavaria are rapidly taking over the economic predominance that was traditionally the prerogative of the Ruhr region. These shifts are partly brought about by the emergence of new, relatively high-quality industries, which are less strongly tied to the traditional industrial areas; attractive residential environments are highly relevant to their selection of a location. The broad pattern of change is reinforced by the healthy growth of the economies of Switzerland and Austria. Also the recent expansion of the EC to incorporate three southern European countries (Greece, Portugal, Spain) underlines the general shift in the European spatial economy.

Clearly, the Netherlands is going to occupy a less central position with respect to the economic center of gravity of Western Europe. Even the construction of the Channel tunnel will reinforce this change. This will not necessarily precipitate the demise of the Dutch economy. Sweden, which has frequently served as a role model for the Netherlands, has been able to maintain a strong economy. However, to preserve its economic vitality in the face of these impending changes, the Netherlands must boost the quality of its industrial output and strengthen its gateway function. The shifts are of great significance for the way in which the 'Gateway to Europe' develops. In order to retain this strong aspect of the Dutch economy, the major international connections must receive more attention. The South will automatically become more central with respect to the important flows of traffic and the economic points of gravity. Conversely, the peripheral position of the North will be emphasized. Undoubtedly, the increasing attention for southern locations among foreign entrepreneurs is an expression of this shift. It will necessitate the adaptation of the traditional way of looking at the Netherlands in terms of center, intermediate zone and periphery.

9.2 Regional Contributions to the National Economy

So far, the regional development chances of industry have been discussed. In preceeding chapters, it was stated time and again that developments in the service sector are quite different from those in industry. Nevertheless, the spatial development of the industrial sector closely reflects that of the economy as a whole. In Chapter 7, for instance, it was discussed how the successful segments of the wholesaling sector are situated predominantly in the central and southern parts of the country. Financial and insurance firms and business services are still strongly tied to the West. However, as can be inferred from Table 9.1, even these sectors are not immune to deconcentration tendencies.

A completely different way of expressing the place and function of the regions, albeit with a similar result, can be derived from the study by Oosterhaven and Stol (1985). They determined the relative contribution of each province

Table 9.1 Employment in the commercial office sector (SIC 8) 1973-1982 (in percentages).

Area	1973	1982	1982 (1973=100)
West	65.7	62.3	137
Center	21.3	24.4	166
Periphery	13.1	13.3	147
Total	100	100	
Absolute total	284,095	410,549	145

Source: Van Dinteren 1986, 267

to the gross national product (GNP); this relative contribution is defined as the absolute share of each province in the GNP minus its share of the national labor force. Because of the weight of their non-industrial sectors, such as services, tourism, etc., a relatively favorable score for the provinces of Noord-Holland and Zuid-Holland is found (Table 9.2). The relatively important contribution of the three northern provinces and of Zeeland are also noteworthy in the table. Even if the contribution made by the extraction of natural gas is omitted, the province of Groningen still contributes a full percentage point more to the GNP than could be expected on the basis of the size of its labor force alone. Furthermore, it is noticable that, in spite of the continuing strong position of Noord-Holland and Zuid-Holland, the erstwhile large differences among the provinces are decreasing. The function of the West as the engine of the Dutch economy is waning.

The level of deconcentration in the Dutch economy can

Table 9.2   Relative contribution of provinces to the gross domestic
            product, in percentages.

Province	Contribution to GDP				
	1960	1965	1970	1975	1981
Groningen	-0.1	0.4	-0.7	+0.2	+1.0
(incl. natural gas)			+0.1	+3.5	+4.8
Friesland	-0.9	-0.5	-0.2	-0.3	-0.1
Drenthe	-0.8	-0.8	-0.1	0.0	+0.5
Overijssel	-1.3	-0.1	-0.7	-0.4	-0.5
Gelderland	-1.3	-1.5	+2.0	-1.1	-1.2
Utrecht	-0.2	-0.5	+1.2	-0.3	-0.4
Noord-Holland	+2.4	+2.0	-0.1	+0.3	+0.7
Zuid-Holland	+3.8	+3.7	+0.1	+1.6	+1.0
Zeeland	-0.3	-0.2	0.0	+0.5	+0.1
Noord-Brabant	-0.7	-0.3	-0.4	-0.2	-0.8
Limburg	-0:6	-0.6	-1.1	-0.3	-0.5

Source: Oosterhaven & Stol 1985, 14

be illuminated by comparing the regional contributions to
the GNP in 1973 and 1982 on the level of the so-called COROP
regions. This analysis shows clearly how a number of large
concentrations of populations as well as some traditional
industrial areas lost ground: the larger Amsterdam region,
the central part of the province of Gelderland, Zuid-
Limburg, Rijnmond, IJmond, and Zeeuws-Vlaanderen. The parts
of the province of Groningen outside the city itself and the
northern part of Drenthe made some gains, but this is
because of the gas extraction. Much more impressive are the
gains made by the southeastern part of the province of
Drenthe, which was the first area recognized as a problem
region and is still part of that league. Gains were also
made by the horticultural areas of Westland and the area to
the north of the city of Leiden, famed for its tulip fields.
Some areas in the central part of the country (Utrecht, De
Veluwe and the central part of the province of Noord-
Brabant) also increased their relative contribution to the
GNP. In the peripheral areas in the North and in the
province of Limburg other than the ones listed here, the
contribution to the GNP was mostly stable.

9.3 Summary

Regional policy is increasingly concerned with the contribution of regions to the national economy. Given the policy to
boost and to strengthen the national economy structurally,
the attention for the production potential of the regions is
essential. Regional policy should aim for more than to find
solutions for acute problems. It should also be concerned

with the use that is made of existing potentials and opportunities. The analysis of the spatial dimensions of the Dutch economy presented above shows clear tendencies toward deconcentration. Outside the West, the intermediate zone contains regions that offer many opportunities. The recent economic recovery, in industry as well as in basic service sectors, is clearly felt in these regions. The North and Limburg, however, may lack clear indications of growth, but at least they have been stopped from slipping further within the national economy.

Bibliographical References

ALDERS, B. C. M. & P. A. DE RUYTER (1984), *De ruimtelijke spreiding van kansrijke ekonomische aktiviteiten in Nederland; vooronderzoek.* Apeldoorn/Delft: STB/TNO.

ANSOFF, H. J. (1965), *Corporate Strategy, an Analytic Approach to Business Policy for Growth and Expansion.* New York: McGraw-Hill.

BARTELS, C. P. A. & J. J. VAN DUIJN (1981), *Regionaal-ekonomisch beleid in Nederli xd.* Assen: Van Gorcum.

BEKKERS, H., B. VAN DIJCK & J. SALVERDA (1985), *Internationale ondernemingen in de regio's Amsterdam en Utrecht. Vestigingstendensen van handels- en dienstverlenende bedrijven en hun regionale effekten.* Utrecht: Geografisch Instituut, Stepro-rapport 30d.

BERG, L. VAN DEN (1986), De informatiemaatschappij: een nieuwe uitdaging voor de grote stad. *Economisch-Statistische Berichten* 71, pp. 925-30.

BIRCH, D. L. (1979), *The Job Generation Process.* Cambridge, Mass.: MIT.

BOS, R. W. J. M. (1976), Van periferie naar centrum; enige kanttekeningen bij de Nederlandse industriële ontwikkeling in de negentiende eeuw. *Maandschrift Economie* 40, pp. 181-205.

BOSCH, L. H. M., W. DRIEHUIS & P. J. VAN DEN NOORD (1985), *Het werkgelegenheidswonder in de Verenigde Staten: mythe of werkelijkheid?* OSA-voorstudies nr. V4. Den Haag: OSA.

BOUMAN, H., T. THUIS & B. VERHOEF (1985), *High-tech in Nederland. Vestigingsplaatsfactoren en ruimtelijke spreiding.* Amsterdam/Utrecht: VU/RUU.

BOUMAN, P. J. (1954), Gedenkboek Wilton-Feijenoord. Schiedam.

BUCK, R. (1985), *Nieuwe Japanse en Amerikaanse industriële vestigingen in West-Europa.* Nijmegen: Geografisch en Planologisch Instituut.

BUREAU ECONOMISCH ONDERZOEK (1983), *Amsterdam computerstad. Verkenning van de computerindustrie en -dienstverlening.* Amsterdam: Gemeente Amsterdam.

DAVELAAR, E. J. & P. NIJKAMP (1986), De stad als broedplaats van nieuwe activiteiten. *Stedebouw en Volkshuisvesting* 71, pp. 61-6.

DAVELAAR, E. J. & P. NIJKAMP (1987), Industriële innovaties en de broedplaatsgedachte van steden. *Economisch-Statistische Berichten* 72, pp. 716-22.

DEURLOO, M. C. & G. A. HOEKVELD (1981), The population growth of the urban municipalities in the Netherlands between 1849 and 1970. In: H. Schmal (ed.), *Patterns of European Urbanization Since 1500.* London: Croom Helm, pp. 247-83.

DIEPERINK, H. & P. NIJKAMP (1986), De agglomeratie-index, *Planning* 27, pp. 2-8.

DINTEREN, J. VAN (1986), Commerciële kantoren en Nederlandse steden. *Geografisch Tijdschrift* XX, pp. 263-72.

DINTEREN, J. VAN (1987), The role of business service offices in the economy of medium sized cities. *Environment and Planning* A 19, pp. 669-86.

DOOREN, A. J. , J. G. DUBBELMAN & J. J. REYNDERS (1985), *Industriële vestigingen van buitenlandse ondernemingen in West en Midden Brabant.* Utrecht: Geografisch Instituut, Stepro-rapport 30c.

DOORN, J. A. A. VAN (1960), *De Nederlandse ontwikkelingsgebieden. Schets van een problematiek.* 's-Gravenhage: Staatsuitgeverij.

DOORN, P. K. (1985), A multivariate classification of the Dutch labour force into socio-cultural, demographic and economic groups. *European Political Data Newsletter* no. 56, pp. 6-17.

DOORN, P. K. (1986), Een nieuwe sociale kaart van Nederland: stratifikatie, mobiliteit en regionale arbeidsmarkten. *Paper NSAV Conference* Amsterdam.

DUIJN, J. J. VAN (1979), *De lange golf in de economie.* Assen: Van Gorcum.

DUIJN, J. J. VAN (1980), Economisch beleid en industriële ontwikkelingsfase. In: W. R. R., *Sectorstructuurbeleid: mogelijkheden en beperkingen.* 's-Gravenhage: Staatsuitgeverij, pp. 55-80.

DUIJN, J. J. VAN (1983), *The Long Wave in Economic Life.* London: Hutchinson.

DUIJN, J. J. VAN (1985), Het veranderende karakter van de investeringen. In: A. van der Zwan (ed.), *Nederland in zaken. Investeren, winst en werkgelegenheid.* Utrecht: Veen, pp. 33-51.

DUNNING, J. H. (1980), Towards an eclectic theory of international production: empirical tests. *Journal of International Business Studies* 11, pp. 9-30.

EVERWIJN, J. C. A. (1912), *Beschrijving van handel en nijverheid in Nederland.* 's-Gravenhage: Belinfante.

FEYTER, C. A. DE (1982), *Industrial Policy and Shipbuilding.* Utrecht: Hes.

FISCHER, A. (1979), Politique régionale et stratégie spatiale de la grande firme, l'exemple de Philips aux Pays-Bas. *Norois* 26, pp. 49-66.

FISCHER, A. (1980), *L'industrialisation contemporaine des Pays-Bas. Recherche sur l'évolution des régions périphériques*. Paris: Sorbonne N.I. Recherches 41.

FRANKE, B.H.V.A. & W.A.C. WHITLAU (eds 1979), *De Vennootschap Nederland*. Deventer: Kluwer.

GALBRAITH, J.K. (1966), *The New Industrial State*. Boston: Houghton Mifflin.

GEERLOF, J. (1983), Toenemende verschillen in regionale werkloosheid. *Economisch-Statistische Berichten* 68, pp. 765-69.

GELDER, W. VAN (1973), *Industriepolitiek*. Amsterdam: WBS.

GEORGE, P. (1961), Les établissements Philips aux Pays-Bas, une politique de la répartition géographique des usines. *Bulletin d'association des géographes Française*, no. 301-302, pp. 198-205.

GERRETSON, C. (1937-1942), *Geschiedenis der Koninklijke*. Utrecht: Kemink, 3 vols.

GIBB, J.M. (ed., 1985), *Science Parks and Innovation Centres: their Economic and Social Impact*. Amsterdam: Elsevier.

GODDARD, J.B. (1975), *Office Location in Urban and Regional Development*. London: Oxford University Press.

GRIFFITHS, R.T. (1979), *Industrial retardation in the Netherlands, 1830-1850*. 's-Gravenhage: Martinus Nijhoff.

GROENEWEGEN, J. (1985), Industriebeleid in Frankrijk. *Maandschrift Economie* 49, pp. 226-40.

HAAS, H. VAN DER (1967), The enterprise in transition. An analysis of American and European practice. Delft, TU (Ph.D.).

HÅKANSON, L. (1979), Towards a theory of location and corporate growth. In: F.E.I. Hamilton & G.J.R. Linge (eds), *Spatial Analysis, Industry and the Industrial Environment*. Chichester: Wiley, Vol. 1, pp. 115-38.

HAMILTON, F.E.I. (1976), Multinational enterprise and the European economic community. *Tijdschrift voor Economische en Sociale Geografie* 67, pp. 258-78.

HART, H.W. TER (1979), *Vestigingsplaatsaspecten van topmanagement*. Meppel: Krips Repro (Ph.D. University of Amsterdam).

HEEREN, H. (1985), *Bevolkingsgroei en bevolkingsbeleid in Nederland*. Amsterdam: Kobra.

HEUVELHOF, E.F. TEN & S. MUSTERD (1983), De 'vruchtbaarheid' van gebieden voor het starten van bedrijven. *Economisch-Statistische Berichten* 68, pp. 799-805.

HIRSCHMANN, A.O. (1958), *The Strategy of Economic Development*. New Haven, Conn.: Yale University Press.

208

HOEK, J. J. VAN DEN (1956), *Industriële vestigingspolitiek van de overheid.* Leiden: Stenfert Kroese.

HOEVEN, P. J. A. TER (1963), *Havenarbeiders van Amsterdam en Rotterdam.* Leiden: Stenfert Kroese.

HOFSTEE, E. W. (1950), *Economische ontwikkeling en bevolkingsspreiding.* 's-Gravenhage: RNP nota nr. 3.

HOOGSTRATEN, P. VAN (1983), *De ontwikkeling van het regionaal beleid in Nederland 1949-1977.* Eindhoven, TU (Ph. D.).

HORVERS, A. & E. WEVER (1987), De alerte onderneming. *Economisch-Statistische Berichten* 72, pp. 795-99.

JANSEN, A. C. M. (1972), Enkele aspekten van het ruimelijk gedrag van grote industriële concerns in Nederland, 1950-1971. *Tijdschrift voor Economische en Sociale Geografie* 63, pp. 411-25.

JANSEN, A. C. M. (1981), 'Inkubatie-milieu': analyse van een geografisch begrip. *Geografisch Tijdschrift* XV, pp. 306-16.

JANSEN, A. C. M., M. DE SMIDT & E. WEVER (1979), *Industrie en ruimte.* Assen: Van Gorcum (Second edition).

JONG, H. W. DE (1980), Het Nederlandse structuurbeleid: ' De zichtbare vinger aan de onzichtbare hand'. In: W. R. R., *Sectorstructuurbeleid: mogelijkheden en beperkingen.* 's-Gravenhage: Staatsuitgeverij, pp. 17-53.

JONG, H. W. DE (1985 a), Industriepolitiek: een lege doos. *Economisch-Statistische Berichten* 70, pp. 192-197.

JONG, H. W. DE (1985 b), De internationalisatie van het Nederlandse bedrijfsleven. In: M. T. Brouwer & H. W. ter Hart (eds), *Ondernemen in Nederland, Mislukkingen en Mogelijkheden.* Deventer: Kluwer pp. 133-152.

JONG, M. W. DE (1983), Regionale condities voor nieuwe hoogwaardige bedrijvigheid. Industriële vernieuwing in Greater Boston. *Economisch-Statistische Berichten* 68, pp. 1059-63.

JONG, Th. J. M. DE & M. J. T. THUIS (1983), *Grote industriële ondernemingen in de Europese Gemeenschap.* Utrecht: Geografisch Instituut.

JONGE, J. A. DE (1968), *De industrialisatie in Nederland tussen 1850 en 1914.* Amsterdam (Ph. D.)/Nijmegen: SUN 1976.

JONGE, J. A. DE (1978), Het economisch leven in Nederland 1873-1895, 1895-1914. In: *Algemene Geschiedenis der Nederlanden,* Vol. 13. Haarlem: Fibula-van Dishoeck, pp. 35-56, 249-84.

KAMANN, D. J. F. (1985), Industrial organization, innovation and employment. In: P. Nijkamp, ed., *Technological Change, Employment and Spatial Dynamics.* Berlin: Springer, pp. 131-54.

KEIZER, D. P. (1985), *Friese vrijheid in ondernemersland. Een studie naar de externe controle in de industrie.* Assen:

Van Gorcum.

KEMPER, N.J. & M. DE SMIDT (1980), Foreign manufacturing establishments in the Netherlands. *Tijdschrift voor Economische en Sociale Geografie* 71, pp. 21-40.

KEUNING, H.J. (1955), *Mozaïek der functies.* Den Haag: Leopold.

KLAASSEN, L.H. & P. DREWE (1973), *Migration Policy in Europe.* Farnborough: Saxon House.

KLEINKNECHT, A. & A. MOUWEN (1985), Regionale innovatie (R&D): verschuiving naar de Halfwegzone? In: W.T.M. Molle (ed.), *Innovatie en regio.* 's-Gravenhage: Staatsuitgeverij, pp. 125-42.

KNAAP, G.A. VAN DER (1978), *A Spatial Analysis of the Evolution of an Urban System: the Case of the Netherlands.* Rotterdam: Erasmus University (Ph.D.).

KOERHUIS, H. & W. CNOSSEN (1982), *De software- en computerservicebedrijven.* Groningen: Geografisch Instituut, Sociaal-Geografische Reeks 23.

KOHNSTAMM, G.A. (1947), *De toekomst van Nederlands industriële ontwikkeling.* 's-Gravenhage: Martinus Nijhoff.

KOJIMA, K. (1977), Direct foreign investment between advanced industrialized countries. *Hitotsubashi Journal of Economics* 18, pp. 1-18.

KOK, J.A.M.M., G.J.D. OFFERMAN & P.H. PELLENBARG (1984), *Innovatieve bedrijven in Nederland. Een onderzoek naar de aard en regionale spreiding van innovaties in het Nederlandse midden- en kleinbedrijf.* Groningen: Geografisch Instituut, Sociaal-Geografische Reeks 32.

KOK, J.A.M.M., G.J.D. OFFERMAN & P.H. PELLENBARG (1985), *Innovatieve bedrijven in het grootstedelijk milieu.* Groningen: Geografisch Instituut, Sociaal-Geografische Reeks 34.

KORTLANDT, A. (1949), *De verspreiding van de bevolking in Nederland.* 's-Gravenhage: R.N.P. publikatie nr. 3.

KRUSEL, G. (1957), *De industriële vestigingsplaats.* Leiden: Stenfert Kroese.

KRUYT, J.P. & W. GODDIJN (1962), Verzuiling en ontzuiling als sociologisch proces. In: A.N.J. den Hollander et al., *Drift en koers.* Assen: Van Gorcum, pp. 227-63.

KUHRY, B. & R. VAN OPSTAL (1988), De arbeidsmarkt naar opleidingscategorie 1975-2000. *Economisch-Statistische Berichten* 73, pp. 72-8.

KUZNETS, S. (1953), *Economic Change.* New York: W.W. Norton.

LAMBOOY, J.G. (1985), *Regionale ontwikkelingen.* 's-Gravenhage: P.R.O. Voorstudie 4.

LEONE, R.A. & H.J. STRUYK (1976), The incubator hypothesis: evidence from five SMSA's. *Urban Studies* 13, pp. 325-31.

LIAGRE BÖHL, H. DE, J. NEKKERS & L. SLOT (eds, 1981), *Nederland industrialiseert! Politieke en ideologiese strijd rondom het naoorlogs industrialisatiebeleid.* Nijmegen: SUN.

LIJPHART, A. (1968), *Verzuiling, pacificatie en kentering in de Nederlandse politiek.* Amsterdam: Elsevier.

LINDEN, J. T. J. M. VAN DER (1985), *Economische ontwikkeling en de rol van de overheid, Nederland 1945-1955.* Amsterdam: Kobra.

LOEVE, A. (1984), *Internationale ondernemingen in Nederland. Omvang en lokatie van vestigingen in Nederland.* Utrecht: Geografisch Instituut, Stepro-rapport 30a.

LOEVE, A. (1985), *Internationale ondernemingen in Nederland. Vestiging van buitenlandse bedrijven in Nederland (1975-1984): motieven en effekten.* Utrecht: Geografisch Instituut, Stepro-rapport 30b.

LOEVE, A. (1986), Buitenlandse ondernemingen in Nederland. *Economisch-Statistische Berichten* 71, pp. 220-25.

McCONNELL, J. (1983), The international location of manufacturing investments: recent behaviour of foreign-owned corporations in the United States. In: F. E. I. Hamilton & G. J. R. Linge (eds), *Spatial Analysis, Industry and the Industrial Environment.* Chichester: Wiley. Vol. 3, pp. 337-58.

McKINSEY & CO. (1978), *Aantrekkelijkheid van Nederland voor buitenlandse investeerders.* Amsterdam.

MANDEL, E. (1968), *Die EWG und die Konkurrenz Europa-Amerika, 'res novae'.* Frankfurt: Europaïsche Verlagsanstalt.

MARE, J. T. DE (1985), De multinationalisatie van de Nederlandse economie. *Economisch-Statistische Berichten* 70, pp. 740-46.

MENSCH, G. (1975), *Das technologische Patt.* Berlin: Umschau Verlag.

MOKYR, J. (1974), The industrial revolution in the Low Countries in the first half of the nineteenth century; a comparative study. *Journal of Economic History* 34, pp. 365-91.

MOLLE, W. T. M. (1983), *Industrial change, innovation and location. Empirical evidence in the Netherlands.* Paper Workshop Research, Technology and Regional Policy. Paris: OECD.

MONHEIM, H. (1972), *Zur Attraktivität deutscher Städte.* München: WGI Berichte zur Regionalforschung 8.

MUNTENDAM, J. (1987), Philips in the World. A view of a mulinational on resource allocation. In: G. A. van der Knaap & E. Wever (eds), *New Technology and Regional Development.* London: Croom Helm, pp. 136-44.

MYRDAL, G. (1957), *Economic Theory and Under-developed Regions.* London: Duckworth.

Nederlands Economisch Instituut - NEI (1984), *Technologische vernieuwing en regionale ontwikkeling.* Rotterdam: NEI.

Nederlandse Middenstands Bank - NMB (1981), *Jaarverslag 1980.* Amsterdam: NMB.

Nederlandse Middenstands Bank - NMB (1985), *De relatie tussen grote en kleine bedrijven in de industrie.* Amsterdam: NMB.

NEWBOULD, G. D., P. J. BUCKLEY & J. THURWELL (1978), *Going International - the Experience of Smaller Companies Overseas.* London: Associated Business Press.

NIEUWKERK, M. VAN & R. P. SPARLING (1985), De internationale investeringspositie van Nederland. *Monetaire Monografie De Nederlandsche Bank,* nr. 4. Deventer: Kluwer.

NOOTEBOOM, B. (1986), De grootheden van de kleintjes. Een overzicht van het midden- en kleinbedrijf. *Economisch-Statistische Berichten* 71, pp. 272-77.

OECD (1981), *Recent International Direct Investment Trends.* Paris: OECD.

OECD (1986), *Netherlands - Economic Surveys 1986/1986.* Paris: OECD.

OOSTERHAVEN, J. & K. STOL (1985), De positie van de provincies in het regionale stimulerings- en ontwikkelings-beleid. *Maandschrift Economie* 49, pp. 3-18.

OSA (1985), *Werk voor allen* 's-Gravenhage: Staats-uitgeverij.

PELLENBARG, P. H. & W. J. MEESTER (1984), Location decisions and spatial cognition. In: M. de Smidt & E. Wever (eds), *A Profile of Dutch Economic Geography.* Assen: Van Gorcum, pp. 105-29.

POEL, H. VAN DER & M. VALKENHOEF (1985), *Nederland in de (uit)verkoop: acquisitie van buitenlandse bedrijven door regionale en lokale overheden.* Utrecht: Geografisch Instituut.

PRED, A. (1977), *City Systems in Advanced Economies.* London: Hutchinson.

RAEDTS, E. P. M. (1974), *De opkomst, de ontwikkeling en de neergang van de steenkolenmijnbouw in Limburg.* Assen: Van Gorcum.

RHIJN, A. A. T. VAN (1971), *De industriële sector en struc-tuurpolitiek.* 's-Gravenhage: Martinus Nijhoff.

ROELOFS, B. & E. WEVER (1985), *Regio en ekonomische poten-tie.* 's-Gravenhage: R. P. D. Studierapport 28.

ROOIJ, P. DE (1979), *Werklozenzorg en werkloosheidsbestrij-ding 1917-1940.* Amsterdam: Van Gennep.

ROSTOW, W. W. (1963), *De vijf fasen van economische groei.* Utrecht: Spectrum.

RUIJTER, P. DE (1978), Broedplaats en reservaat; een onder-zoek naar bedrijven in hun begin- of eindfase in de binnenstadsrand van Zwolle. *Bijdragen tot de Sociale Geografie* 12. Amsterdam: Vrije Universiteit.

SCHENK, E. J. J. (1985), Industriepolitiek: een ' common sense' benadering. *Maandschrift Economie* 49, pp. 192-213.

SCHOUTEN, C. W. (1984), De lange golf. 's-Gravenhage: SMO.

SCHRÖDER, M., M. DE SMIDT & W. R. STARING (1984), Corporate growth, foreign investment and locational choice. In: M. de Smidt & E. Wever (eds), *A Profile of Dutch Economic Geography.* Assen: Van Gorcum, pp. 63-84.

SCHUMPETER, J. A. (1939), *Business Cycles.* New York: McGraw Hill, 2 Vols.

SERVAN-SCHREIBER, J. J. (1967), *Le défi américain.* Paris: Denoël.

SHAPERO, A. (1980), *The Entrepreneur, the Small Firm and Possible Policies.* Limerick.

SHAPERO, A. (1983), *New Business Formation.* Conference BTC, Enschede.

SMIDT, M. DE (1966), Foreign industrial establishmens located in the Netherlands. *Tijdschrift voor Economische en Sociale Geografie* 57, pp. 1-19.

SMIDT, M. DE (1967), Stuwend en verzorgend, een verkenning van de ontwikkeling der konceptie. *Bulletin Geografisch Instituut R. U.* Utrecht, nr. 4, pp. 7-40.

SMIDT, M. DE (1973), Stuwende bedrijvigheid in West- en Zuidwest-Nederland. In: W. Steigenga et al., *West-Nederland, chaotische planning of geplande chaos.* Assen: Van Gorcum, pp. 147-80.

SMIDT, M. DE (1981), Innovatie, industriebeleid en regionale ontwikkeling. *Geografisch Tijdschrift* XV, pp. 228-38.

SMIDT, M. DE (1985a), *Bedrijfshuisvesting op lange strek. Lange golf, innovaties en stedelijke cyclus.* Plan 16, pp. 36-41.

SMIDT, M. DE (1985b), Relocation of government services in the Netherlands. *Tijdschrift voor Economische en Sociale Geografie* 76, pp. 232-36.

SMIDT, M. DE (1985c), An Integrated Structure Plan for the Northern Netherlands: A Test of Integrated Regional Planning. In: A. K. Dutt & F. J. Costa (eds), *Public Planning in the Netherlands.* Oxford: Oxford University Press, pp. 122-40.

SMIDT, M. DE, H. F. L. OTTENS & H. A. PLOEGER (1986), Werkgelegenheidsdynamiek in de vier grootste stadsgewesten, in het bijzonder het Utrechtse. In: F. M. Dieleman, A. W. P. Jansen & M. de Smidt (eds). *Metamorfose van de Stad.* Amsterdam/Utrecht: Nederlandse Geografische Studies 19, pp. 19-34.

SMIDT, M. DE & E. WEVER (eds, 1984), *A Profile of Dutch Economic Geography.* Assen: Van Gorcum.

SMITH, I. J. (1985), Foreign direct investment and disinvestment trends in industrialised countries. In: M. Pacione (ed.), *Progress in Industrial Geography.* London: Croom Helm, pp. 142-73.

SOETE, L. (1987), Economische aspecten van technologische verandering. *Economisch-Statistische Berichten* 72, pp. 464-68.

STEIGENGA, W. (1949), Aantekeningen betreffende de Nederlandse werkloosheidsgebieden. *Tijdschrift voor Economische en Sociale Geografie* 40, pp. 81-100.

STEIGENGA, W. (1955), A comparative analysis and a classification of Netherlands towns. *Tijdschrift voor Economische en Sociale Geografie* 46, pp. 105-19.

STEIGENGA, W. (1958), De decentralisatie van de Nederlandse industrie, een economisch-geografische en statistische analyse van de veranderingen in het spreidingspatroon der industriële werkgelegenheid in de periode 1930-1950. *Tijdschrift voor Economische en Sociale Geografie* 49, pp. 129-48. 1

STEIGENGA, W. , STEIGENGA-KOUWE, S. E. & A. A. VAN AMERINGEN (1955), *Bevolkingsgroei en maatschappelijke verantwoordelijkheid*. Amsterdam: De Arbeiderspers.

STERKENBURG, J. J. (1938), Gloeilampen en radio-industrie. In: P. Lieftinck (ed.). *Het bedrijfsleven tijdens de regering van H. M. Koningin Wilhelmina 1898-1938*. Amsterdam, pp. 170-80.

STOKMAN, C. T. M. (1986), *Opkomst van de Halfwegzone: een kansrijke regio*. Amsterdam: E. G. I.

STOLZENBERG, R. (1984), *Het sociaal-wetenschappelijk onderzoek bij het uitbreidings- en structuurplan. Een sociografische opgave*. Eindhoven (Ph. D. Amsterdam).

STOPFORD, J. M. & J. H. DUNNING (1983), *Multinationals: Company Performance and Global Trends*. London: MacMillan.

STRUYK, R. J. & F. J. JAMES (1975), *Intra-metrolopitan Industrial Location*. Lexington: Lexington Books.

STUIJVENBERG, J. H. VAN (ed. , 1978), *Economische geschiedenis van Nederland*. Groningen: Wolters-Noordhoff.

STULEMEIER, C. (1938), De technische en economische ontwikkeling der kunstzijde-industrie. In: P. Lieftinck (ed.), *Het bedrijfslevens tijdens de regering van H. M. Koningin Wilhelmina 1898-1938*. Amsterdam, pp. 218-32.

STUURMAN, S. (1983), *Verzuiling, kapitalisme en patriarchaat*. Nijmegen: SUN.

TAYLOR, M. (1987), Enterprise and the product-cycle model: conceptual ambiguities. In: G. A. van der Knaap & E. Wever (eds), *New Technology and Regional Development*. London: Croom Helm, pp. 75-94.

TÖRNQVIST, G. (1970), *Contact Systems and Regional Development*. Lund: Gleerup.

TROMP, T. P. (1958), *Les usines Philips et la décentralisation industrielle*. Cahiers de Bruges no. 2.

VANHOVE, N. D. (1962), *De doelmatigheid van het regionaal-economisch beleid in Nederland*. Gent-Hilversum: Brand.

VANNESTE, O. (1967), *Het groeipoolconcept en de regionaal-*

    economische politiek; toepassing op de Westvlaamse economie. Antwerpen: Standaard.

VERHOEVEN, W. H. J. & J. G. VIANEN (1984), *Het midden- en kleinbedrijf. Kansen voor herindustrialisatie.* 's-Gravenhage: EIM.

VERMOOTEN, W. H. (1949), *Stad en land in Nederland en het probleem der industrialisatie.* Amsterdam: Paris.

VERNON, R. (1971), *Sovereignty at Bay: The Multinational Spread of US Enterprises.* Hammondsworth: Penguin.

VLESSERT, H. H. & C. P. A. BARTELS (1985), *Kenniscentra als elementen van het regionale produktiemilieu.* Oudemolen: Buro Bartels.

VRIES, J. DE (1968), *Hoogovens IJmuiden 1918-1968, ontstaan en groei van een basisindustrie.* IJmuiden: K. N. H. S.

VRIES, J. DE (1977), *De Nederlandse economie tijdens de 20ste eeuw.* Haarlem: Fibula.

WATTS, H. D. (1981), *The Branch Plant Economy.* London: Longmans.

WEBBINK, A. H. (1984), *De rol van het midden- en kleinbedrijf in de industrie.* Zoetermeer: EIM.

WEBBINK, A. H. (1985a), *Groot en klein in de industrie, een onderzoek naar groottestructuur en levenscyclus in de Nederlandse industrie.* Zoetermeer: EIM.

WEBBINK, A. H. (1985b), *Innovatie en het MKB.* Zoetermeer: EIM.

WEBER, A. (1909). *Über den Standort der Industrie: 1. Teil. Reine Theorie des Standorts.* Tübingen: Mohr.

WEISGLAS, M. (ed., 1952), *De Nederlandse industrie sinds 1945. Een wereld van groei.* 's-Gravenhage: Sijthoff.

WEMELSFELDER, J. (1985), Kan het ontstaan van nieuwe technologieën worden beïnvloed? *Economisch-Statistische Berichten* 70, pp. 220-24.

WEVER, E. (1971), *Enkele aspecten van industriële ontwikkeling in Nederland tussen 1950 en 1963.* Nijmeegse Geografische Cahiers no. 1. Nijmegen.

WEVER, E. (1974), *Olieraffinaderij en petrochemische industrie. Ontstaan, samenstelling, voorkomen van petrochemische complexen.* Groningen (Ph. D.).

WEVER, E. (1981), Plaats en toekomst van de Nederlandse industrie, *Geografisch Tijdschrift* XV, pp. 218-28.

WEVER, E. (1984), *Nieuwe bedrijven in Nederland.* Assen: Van Gorcum.

WEVER, E. (1986), New firm formation in the Netherlands. In: D. Keeble & E. Wever (eds), *New Firm Formation and Regional Development in Europe.* London: Croom Helm, pp. 54-75.

WEVER, E. (1987), The spatial pattern of high-growth activities in the Netherlands. In: G. A. van der Knaap & E. Wever (eds), *New Technology and Regional Development.* London: Croom Helm, pp. 165-86.

WEVER, E. & S. GRIT (1984), Research methods and regional
policy. In: M. de Smidt & E. Wever (eds), *A Profile of
Dutch Economic Geography*. Assen: Van Gorcum, pp. 39-62.

WEVER, E. & H. W. TER HART (1986), *Poort van Europa, Atlas
van Nederland*, Vol. II. 's-Gravenhage: Staatsuitgeverij.

WIJERS, G. J. (1982), *Industriepolitiek. Een onderzoek naar
de vormgeving van het overheidsbeleid gericht op industriële sectoren*. Leiden: Stenfert Kroese.

WIJERS, G. J. (1985), Een pleidooi voor industriepolitiek.
Maandschrift Economie 49, pp. 175-91.

WIJKSTRA, R. (1979), Herstructurering van de Nederlandse
industrie en internationale arbeidsverdeling. In:
S. L. Kwee, J. G. Lambooy, J. Buit & M. de Smidt (eds),
Nederland op weg naar een post-industriële samenleving?
Assen: Van Gorcum, pp. 57-63.

WILSON, C. (1954), *Geschiedenis van Unilever*. 's-Gravenhage:
Martinus Nijhoff.

WINDMULLER (1969), *Labor Relations in the Netherlands*.
Ithaca N. Y.

WINSEMIUS, J. (1945/49), *Vestigingstendensen van de Nederlandse nijverheid*. 's-Gravenhage: Staatsuitgeverij.

WINSEMIUS, J. (1952), Industrialisatie van de zogenaamde
ontwikkelingsgebieden in Nederland. *Tijdschrift van het
Kon. Ned. Aardr. Gen.* LXIX, pp. 3-15.

WOLFF, L. DE (1983), Het geheim van succesvol ondernemen.
Intermediair 19 (48), pp. 21-41.

W. R. R. (1980), *Plaats en toekomst van de Nederlandse industrie*. 's-Gravenhage: Staatsuitgeverij.

ZWAN, A. VAN DER (1980), Inleiding. In: W. R. R. Sectorstructuurbeleid: *Mogelijkheden en beperkingen*.
's-Gravenhage: Staatsuitgeverij, pp. 1-15.

Index